PRECIOUS

www.penguin.co.uk

PRECIOUS

The History and Mystery
of Gems Across Time

HELEN MOLESWORTH

doubleday

TRANSWORLD PUBLISHERS
Penguin Random House, One Embassy Gardens,
8 Viaduct Gardens, London SW11 7BW
www.penguin.co.uk

Transworld is part of the Penguin Random House group of companies
whose addresses can be found at global.penguinrandomhouse.com

Penguin
Random House
UK

First published in Great Britain in 2024 by Doubleday
an imprint of Transworld Publishers

A CIP catalogue record for this book
is available from the British Library.

ISBN 9780857529091

Typeset in 11.75/16pt Minion Pro by Jouve (UK), Milton Keynes
Printed and bound in Great Britain by Clays Ltd, Elcograf S.p.A.

The authorized representative in the EEA is Penguin Random House Ireland,
Morrison Chambers, 32 Nassau Street, Dublin D02 YH68.

Penguin Random House is committed to a sustainable
future for our business, our readers and our planet. This book is made from
Forest Stewardship Council® certified paper.

To Mum and Dad, the most
precious gems of all.

Contents

KEY ANCIENT GEM SOURCES

 1. EMERALD
  Egypt, Pakistan,
  Austria

 2. RUBY
  Sri Lanka

 3. SAPPHIRE
  Sri Lanka

 4. GARNET
  Sri Lanka, India,
  Czech Republic,
  Sweden, Portugal,
  Turkey, Sinai Peninsula

 5. PEARL
  Persian Gulf, Red Sea, Gulf
  of Mannar (Sri Lanka),
  Britain (Scotland), China

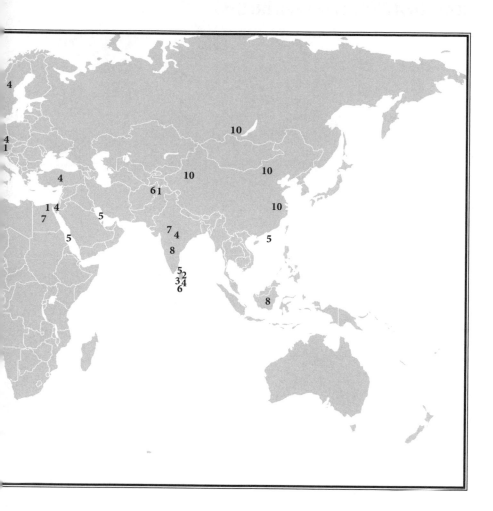

 6. SPINEL

Tajikistan/Afghanistan
(Badakhshan), Sri Lanka

 7. QUARTZ

Egypt, India (and deposits
between the two)

 8. DIAMOND

India, Borneo

 9. COLOURED DIAMOND

None

 10. JADE

China, Russia,
Guatemala, Switzerland

KEY MODERN GEM SOURCES

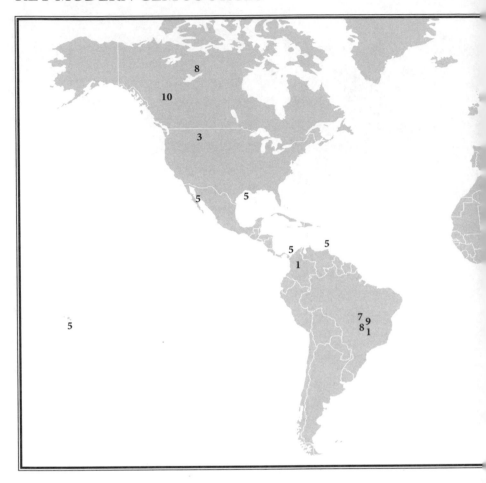

 1. EMERALD

Colombia, Brazil, Zambia, Ethiopia, Afghanistan, Russia, Madagascar, Zimbabwe

 2. RUBY

Burma (Myanmar), Mozambique, Tajikistan, Thailand/Cambodia, Vietnam, Madagascar

 3. SAPPHIRE

Sri Lanka, Burma (Myanmar), Kashmir, Madagascar, Thailand/Cambodia, Vietnam, Tanzania, Australia, USA (Montana)

 4. GARNET

Tanzania, Czech Republic, Kenya (tsavorite), Russia (demantoid)

 5. PEARL

Persian Gulf, Red Sea, Gulf of Mannar (Sri Lanka), Burma (Myanmar), Philippines, Indonesia, Australia, China, Japan, French Polynesia, Gulf of California, Gulf of Mexico, Panama, Venezuela

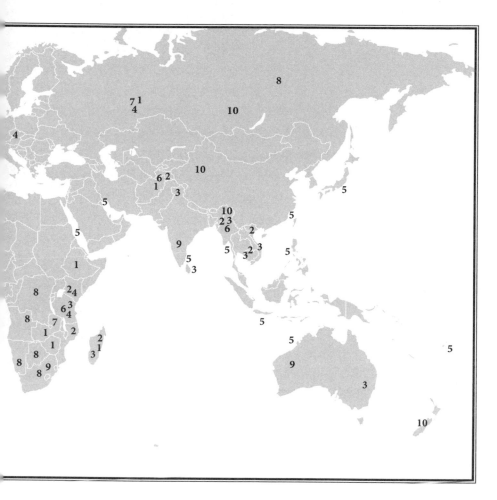

 6. **SPINEL**

Burma (Myanmar), Tanzania,
Tajikistan/Afghanistan

7. **QUARTZ**

Brazil, Russia, Zambia,
Germany

 8. **DIAMOND**

South Africa, Namibia,
Botswana, Angola, Congo,
Brazil, Russia, Canada

 9. **COLOURED DIAMOND**

Australia, South Africa,
Brazil, India

 10. **JADE**

Burma (Myanmar),
China, Canada, Russia,
New Zealand

Introduction

THE MOMENT the box is opened is one every jewellery lover lives for. You take in the faint click of the mechanism and the soft creak of an antique hinge. The smell of the leather. And then the surprise of what has been hiding inside, the sharp glint of a gem that light first touched decades or even centuries ago.

In my life studying, valuing, curating and most of all loving gemstones, I have often been thrilled by moments such as this, not knowing what I am about to see but having a hunch that it will be special. Working in the auction business, you can be sent to value a collection with little more than a name and a blank sheet of paper. Sometimes your eyes are the first in years to appreciate a remarkable jewel that has been living behind the walls of a safe. This joy, of meeting an object you instantly recognize, or immediately clocking the value of something unheralded, is one of the many reasons to relish the work I have been lucky enough to call a career.

That career was one I had never considered before it started. And it might never have come to pass but for an intervention from my father. Having just finished my Classics degree and decided that I didn't want to follow the corporate route, nor to pursue my interest in archaeology professionally, I returned home at a loss. I told him I had no idea what to do with my life, except that I wanted to be happy. His response was a burst of unconventional careers advice, counting the options off on his fingers.

'Hair, make-up, clothes, jewellery. Pick one.'

Far from being flippant, my father had struck a chord. While I had never considered what following my dreams might actually mean, the moment he mentioned jewellery I felt a tingle. One thing I knew for certain was that I had loved gemstones for almost my entire life.

Aged two, a white-haired, blue-eyed baby in my birth country of Kenya, early photographs show me bedecked only in my mother's beads and (matching, I'm glad to say) high heels. I was otherwise 'tupu-tupu' – Swahili for stark naked. I was, I'm told, obsessed with anything sparkly, regularly decorating my toy rocking camel (our equivalent of a rocking horse) with whatever beads and bandeaus I could lay my hands on. When we came to England, I became a tomboy: climbing trees, scrapping with boys and digging stuff up at every opportunity. Yet the obsession with everything that glimmered remained, and I would rush off to local jewellery shops whenever I could with all the pocket money I had scraped together.

So great was my desire for gems that, when I fell in love for the first time at the age of six, it was not with a boy but an amethyst geode: a mini crystal cave of piercingly purple gems that shone with seemingly impossible sharpness and depth. Unable to take my eyes off this treasure, I believed that destiny had brought the most beautiful thing in the world to a back garden in south-west England and placed its spiky symmetry snugly into the palm of my hand.

It was my first glimpse of perfection, hence my dismay when its owner – my godfather – gently told me that it couldn't be mine. I watched him place it out of my reach six foot high on a drystone wall at the end of the garden and wander back to the table. As soon as adult backs were turned, I ran to the wall and started my ascent. Before I could reach the top and claim my prize, my sneaky scrambling brought the whole thing down on me, trapping and breaking my leg. It was an early lesson in the irresistible attraction of gemstones, the lengths that people will go to pursue them, and the danger that can accompany these most exciting of objects. Not to mention the disappointment that is a perennial fellow traveller on the quest for gem perfection. I never did get to keep the amethyst.

A seed had been planted that day, and my interest continued to grow. Later, I found myself integrating the study of ancient jewellery into the tail-end of my Classics degree. It was my archaeology tutor who introduced me to Roman gems: tiny, precious, yet often overlooked ancient works of art. Not long afterwards I was having that offhand conversation about a choice of career in my parents' kitchen, and within weeks I was on Bond Street, knocking on the doors of jewellery houses with a CV in my hand, as green as the emeralds on display. Soon I was having the pages pushed back at me over the table, by a man who very firmly insisted – twice – that he was 'not looking for a secretary'. Another dealer, who I ended up working for, initially told me to forget it: I was the wrong gender, wrong family, wrong religion. Why didn't I just become a lawyer?

It was my first experience of the rough edges that surround an industry dealing in some of the world's hardest objects. But it was also a suitably fiery introduction to a world full of bright and brilliant human gems: people driven by a passion for their subject whose sparks I had already felt igniting inside me. Another knock on a Bond Street door resulted in a much friendlier conversation, culminating with a visit to the safe and my first experience of the treasures of the trade. I still remember the jewels I saw that day. One diamond dealer showed me a hundred-carat stone on the wheel as it was being polished by one of the last diamond cutters left in London. This was a world I knew I wanted to be part of from the moment its door began to creak ajar. I signed up for night classes in the science of gemmology, and so began the greatest love affair of my life: the study of gemstones.

In joining the gem and jewellery industry, I became part of one of the oldest and richest human fascinations. For as long as people have known about gemstones, they have treasured them. They are objects of great beauty, sources of deep symbolism, stores of significant value, subjects of fierce competition, and catalysts of trade and discovery. Gems have a value that spans the aesthetic, cultural, financial and historical: they are significant to the human narrative in almost as many different ways as they can boast sparkling facets. They are the ultimate meeting of science and stories.

The gemstone that shines in the auction room and ignites a bidding war is merely the tip of the iceberg: the glittering culmination of an extraordinary journey through time. That stone is a product of nature's extremes and expanses, born from the collisions and eruptions that shaped the planet as we know it today, as mountains formed and continents closed over long-forgotten oceans. Its natural form was forged at the meeting point of numerous chance happenings: violent geological events which brought together minerals that would never normally meet, provided the necessary heat and pressure for the crystals to form, and raised them from deep below the earth's surface – tens or even hundreds of kilometres down – to within reach of human hands and tools. And its stunning beauty was honed by human hands, relying on centuries of accumulated knowledge of how to cut and polish the rough crystal into a glimmering masterpiece that will set hearts racing.

If holding a truly magnificent gemstone in your hand feels like a miracle, then that is barely an overstatement: a stone that may sell for tens of millions of dollars has endured a journey that may stretch back tens of millions of years. The extent of fate and fortune required to place it on your finger or around your neck is almost impossible to grasp.

By contrast, the hard-earned beauty of a polished gemstone is deceptively straightforward: the glorious gift of light and colour. The way in which gems interact with light is simply incomparable. The crystalline structure of a diamond can do things to light that should bend your mind as much as the wavelengths. A light wave enters the gem through the many facets the cutter has painstakingly polished onto its surface. It dances around within the stone, and in some cases disperses into its rainbow of component colours. If the stone has been cut properly, every single facet will reflect all of that wonder back into the eye of the observer.

The effect can be almost spiritual. Talking, one day, with a friend (who was Buddhist) about why we loved gems so much, I tried to express the way it made me feel to escape into the soul of a stone, examining it for every detail, hunting down information I could use to determine more about where it had originated and what had happened to it; but especially how I found the rest of the world drift away as I did so, a moment of hyperfocus and peace. She replied, so simply and profoundly, 'It's a bit like meditation, isn't it?'

In many gemstones, the power of colour is equally important. Caused by complex chemistry and chance physical anomalies in the crystal, these hues are not just natural phenomena in their own right, but psychologically important, conveying timeless meaning and symbolism. The associations of blue stones with heavenly might, reds with fire, blood and passion, and greens with the colours of nature and rebirth, stretch back millennia and remain resonant today. They help to explain why gemstones have been esteemed not just as aesthetic items but objects that denote power and status, convey religious meaning, and have often been thought to serve a medical purpose.

This is part of what makes the study of gemmology and gem history so fascinating. The flashing fire of the brilliant-cut diamond, the dark pools of the blood-red garnet, and the soothing green of the Colombian emerald do not just reflect and disperse light, engaging and enticing the eye. They also mark a trail of cultural and social significance whose footsteps can be followed from the modern world far back into antiquity.

This book will follow that trail and my own travels down it during nearly twenty-five years in the industry, crossing the world in search of gemstones and their stories – from the mines where they are extracted to the markets in which they are traded, the workshops where they are fashioned, the auction houses where they are tussled over, and the museums and country houses where they come to rest. It will explore the meaning, mystery and history of ten of the most famous and historic gemstones. Some are closely related whereas others are entirely separate from one another, but all are resonant of how gems have adorned human history.

Central to that story is how gemstones provide a point of connection from the present to a long-distant past, allowing us to compare the way we think, the beliefs we hold and the aesthetics we admire to those who lived centuries or millennia before us. So often the earliest history is something we can only squint at through fragmentary pieces of literary evidence or archaeological ruins that give just a hint of the world these civilizations inhabited. But the innate durability of gems, often preserved in the graves of those who owned them, or in the hoards of those who stashed them, means they can survive intact, as if they had not aged a day in over a thousand years. Like us, our ancestors wore these precious objects close to their skin, and close to their hearts. With a piece of jewellery that once lived around the neck of a Roman aristocrat or on the hip of an Anglo-Saxon royal, we can hold that history in our palm: seeing what they saw, feeling what they felt, and even sensing the faint shadow of their presence.

Gemstones also open a wider window into history, with their stories helping to reveal much about the past that might otherwise remain hidden. They illuminate the nature of belief and superstition in ancient Egypt, Rome and Greece; the dynamics of power and status, commercial trade and cultural exchange in the early medieval world once demeaned as the Dark Ages; the ambition of dynasty building in Mughal India and Napoleonic France; and the reality of how colonial expansion drove the extraction and exploitation of gemstones, from South America in the sixteenth century to Africa in the nineteenth and twentieth centuries.

Gemstones have been a consistent feature of human history, and the history of gems is ultimately a story of people: hands that have sought and found, cut and polished, treated and tested, bought and sold, owned and cherished these remarkable treasures. That human dimension helps to explain why no gemstone group has had an entirely straightforward journey through their long history. All the gemstones this book will explore have seen their fortunes fluctuate, whether because their supply has waxed and waned, their reputation has risen or fallen, or their popularity has not travelled from one place or point in history to another.

As we will see, much of what we now take for granted about gemstones has not always been true: the diamond's clear-cut status as the symbol of eternal love is a relatively modern development, in part shaped by the advertising industry; the ruby was not always the peerless prince of red stones; and the preferred gem of the Mughal emperors, perhaps history's most lavish jewellery collectors, is one that has become less well known and frequently misunderstood – the spinel. Gemstones may be a constant in history, but the way they have been used, prized, valued and marketed has also been the subject of constant change and evolution. These are dynamic and ever-changing objects as well as perennial ones.

I have made a life and career in gemstones not just because I love the objects. It is also a line of work that feels like no other, almost a perfect synthesis of every subject under the sun. The study of gems involves history and politics, it includes archaeology and engineering, geography and geology, chemistry and physics, psychology and romance, fine art and high finance. It is a subject with something for everyone: whether interested in the geology of how gem deposits are formed, the chemistry and physics that lie behind their existence, the craftsmanship that reveals their beauty, the money and markets of prices rising and auction gavels falling, or the simple romance and psychology driving our attraction to their glittering facets and rich colours. In my career, I have worked with miners and geologists, laboratory technicians and scientific researchers, archaeologists and curators, sales executives and auctioneers, stone cutters and valuers, and held several of these roles myself.

This work has taken me all over the world: from the sapphire-rich rivers of Sri Lanka to the auction houses of London and Geneva; from knee-deep snowdrifts in Moscow to 100ft-deep mining shafts in the emerald districts of Colombia. In the gem trade, your work can be at an oligarch's dinner party or in the African bush, wearing an evening dress one week and hard hat the next. Some days will be spent getting lost in a museum archive, and others presenting to high-net-worth clients. I adore the contrast between the earthy and the exquisite, the aesthetic and the scientific, and the gentle rhythm of research set against the sharp tension of an exciting sale. No other work could have allowed me to indulge so many interests: to dig through the dirt as often as I get to try on tiaras.

But nothing has ever quite matched a highlight that came quite early in my career. After the street-level apprenticeship, my jewellery life began in earnest on the graduate scheme at Sotheby's. Within a year of starting I had been sent from London to Geneva to work on multi-million dollar sales, and two years later I was in charge of my first auction back in London. A few years later, not long after I had moved to work at Christie's, the phone in my office rang, with the head of the jewellery department on the line.

'Helen, there's a valuation. You're good at research, I want you to do it.'

Only when I was instructed to present myself at Kensington Palace did I get butterflies. When I arrived the next day with a colleague, we were shown to Princess Margaret's old apartments and greeted by our clients. First a bottle of vintage champagne was opened, and then box after box of her jewellery – from famous pieces she had worn to royal weddings and functions, to sentimental brooches and rings that had never seen the public eye.

Being present at such an appraisal means bearing witness as

tiny, exquisite doorways are opened into history, its artefacts peeking out after years or even decades spent in hiding. And rediscovering stories that might never have otherwise been told. We were handling objects that defined the Princess as a style icon of her era, but also jewels that told of the person and personality, on a private, almost intimate, level. Many of the boxes we opened contained handwritten notes, including one that accompanied a simple sapphire brooch: 'To darling Margaret on her confirmation from her loving Granny Mary. God bless you.' A gift from a Queen to a Princess, but also a simple token of love from a doting grandmother.

The collection ranged from sentimental to spectacular, and one jewel in particular overshadowed them all: the star lot of the sale, the Poltimore Tiara. An intricate canopy of scrolling diamonds, this was without question the iconic piece of Princess Margaret's jewellery. It was the tiara she had worn for her wedding to Antony Armstrong-Jones in 1960, one that wrapped around her famous beehive hair as if it had been made to do so.* The jewel gained fresh renown much later, when a photograph of the Princess wearing it in the bath was published, a scene recreated in *The Crown*. I loved the Poltimore not just as a gorgeous and glorious jewel, but for what it said about this famously independent woman, the royal with a rebellious streak. This was not a piece borrowed from the Royal Collection, as one might expect for the sister of the sovereign, but one she had bought herself, second-hand, at auction. She had even worn it, contrary to tradition, before she was married. This was the tiara not just of a powerful and beautiful

---

* It had in fact been made in the nineteenth century for Lady Poltimore, who had worn it to the coronation of Princess Margaret's grandfather, King George V, in 1911.

woman, but one who knew exactly what she wanted – even if, infamously, she was not always permitted to have it.

It is also a highly versatile item, capable of being worn not only as a diadem but as a necklace, or further disassembled into a series of brooches. And it is the only tiara I have handled in my career that I was specifically forbidden from trying on. But if I couldn't wear the Poltimore, I certainly could get my hands on it – exploring the many forms that had helped make it famous. One of the first things I noticed on taking it out of the box was the screw fittings at the back, and the screwdriver that had been stored underneath it. So I happily got to work one afternoon, as we readied the collection for sale in 2006. It was only when its various pieces were spread on the desk around me that I realized how much time had passed: there was just half an hour before I was due to meet a prominent journalist for a preview and photo shoot. Only some of the quicker handiwork of my career ensured that the collection's signature item was fully present and correct for its first public inspection.

The sale of Princess Margaret's jewellery collection was one of many that have captured my imagination over a career working with gemstones, spanning researching, teaching, curating, sourcing and, of course, selling. These gems and jewels are such a perpetual source of interest because of the stories they tell: the efforts of those who mined and fashioned them; the intimate, personal histories of the individuals and families that owned them; and the sweeping narratives of empires, trade routes, conflict and craftmanship that have made them such an intrinsic part of culture and civilization for millennia.

As objects that have been so consistently and comprehensively loved through human history, they have abundant truths to tell and secrets to share. Gems speak a universal language of human belief and behaviour through history: they illuminate a treasure map of what people did and thought thousands of years ago, and why in so many cases we continue to do and think the same things today. Far more than being objects to look at, they have so much to teach us: a shortcut into whole swathes of human history, beckoning us to understand all the cultures that existed before us, and how close the connections are with our world today.

1

Emerald

The Gem of the Ancients

'Fair speech is more rare than the emerald that is found by slave-maidens among the pebbles.'

THE INSTRUCTION OF PTAH-HOTEP, 2,500 BC

A S THE PICKAXE went through the cellar floor, the workmen paused. It was 1912, East London, and they had a job to do. They were taking down a centuries-old building in such a state of disrepair that it needed razing to the ground and rebuilding from the foundations up. But as they broke through the ancient flooring, a reflection caught the light. There was a glint of something peeking out of a broken wooden box stashed beneath the chalk floor; something shining. They were about to unearth a treasure chest which would astound the world, but which would raise many questions that still remain unanswered. A priceless cache of late sixteenth- and early seventeenth-century jewellery had lain hidden undetected for centuries. It would turn out to be one of history's most significant gem discoveries, and would become known as the Cheapside Hoard.[1]

Now, almost exactly one hundred years later, I had it in front of me. Select pieces had been laid out in preparation for an exhibition to be put on to celebrate the centenary of its discovery, and I was here to examine them. I felt like the proverbial kid in a candy store. There was one jewel which immediately jumped out at me: an enormous emerald crystal, which also enclosed a secret. Inside it was a watch.

Every gemstone asks a question, and the best contain many: puzzles of history and mysteries of time as enticing as the colours and reflections that have long bewitched those who work with and collect them.

But before the questions, before any thought can be given to every hand that touched it on its way to yours – hands that mined it, sculpted it, traded and treasured it – the first sight of a remarkable gemstone prompts something simpler. There is a moment of wonder, taking in something so astonishingly beautiful that it clears the mind, a meditative respite from the questions that will soon follow. This first contact is not analytical, or even professional, but purely emotional. There is only the hypnotizing quality of the colours, the movement of light, and the swirling patterns of the gem's internal world. This admiration of natural beauty is intensified by the technical skill of the artisan, the gem formed and fashioned to bring out a life previously hidden within, and set in a marvel of handiwork, at times as stunning as the stone itself.

The Cheapside Hoard watch is just such a combination: nature's miracle uplifted by supreme human craftsmanship. It is also an unusual gem, in that nature provided the casing for the expert's craft, and not the other way around. A timepiece dating to around 1600 had been embedded in a huge hexagonal emerald crystal several centimetres deep, with a hinged lid, probably cut from the same crystal. The watch face had been cleverly applied with green enamel to blend in, giving the impression of a never-ending gem, and the lid was so fine and transparent that the time could be seen even with the case closed. Although as an object it seemed small – especially sitting in the palm of my hand – the significance of it being made from one single emerald crystal was huge.

The watch is an object whose integrity is as extraordinary as its story. It emerged from a collection of five hundred pieces of predominantly Elizabethan and Jacobean jewellery that lay hidden for centuries beneath a building belonging to the Goldsmith's Company in Cheapside, likely buried by one of its tenant jewellers, and somehow lost and forgotten amid the chaos of one of the great mid-seventeenth-century cataclysms – the English Civil War or the Great Fire of London. Dating to the turn of the seventeenth century, the Hoard is the world's largest collection of original jewellery from this era.[2] Emerald pieces are to the fore, including an exquisite hat ornament in the shape of a salamander, green stones running along its back, diamonds for its eyes and splayed golden feet. Yet emerald is just one of a wide range of stones from diverse origins that decorate the collection: diamonds from India, rubies from Burma, sapphires from Sri Lanka, pearls from the Persian Gulf and peridots from Egypt – all helping to sketch a picture of the continent-crossing supply chains that fed the Elizabethan jewellery trade.

Even among this rich cache, the watch stands out. Unlike most of the other gems in the collection, brought to Europe along the Asian Silk Route, the emerald had an even more exotic origin: it had arrived in Europe from the 'New World', direct from Colombia. It was the product of the bloody Spanish conquest that had been unfolding since the 1530s, in pursuit of the land of

El Dorado.* The Cheapside emerald watch would have been one of the earliest emeralds brought to Europe from Colombia, and one of the best.

That the crystal was from one of the few Colombian emerald mines in operation when Spanish conquistadors arrived in the sixteenth century was confirmed by analysis, but it is also immediately evident from the distinctive size and quality of the crystal. Until the Spanish returned from this 'New World', no large, high-quality emeralds were in use in the West, which relied entirely on small, lower-grade deposits in Egypt, Austria and Pakistan.[3] For the period, the size of the crystal was remarkable. It was also a pristine example of the emerald's natural hexagonal form, polished down to perfection, and given delicate bevelled edges. Although Colombian emeralds were already entering the European market by the mid-1500s, local jewellers would have seen little like this one before.[4] The emerald must have been a near-unique treasure arriving in London, and would have astonished anyone who set eyes on it. The gem certainly left me stunned when I first saw it and picked it up, some four hundred years later.

These clues establish the bare bones of the jewel's origin: an emerald mined in Colombia, transported to Europe by Spanish traders around 1600, and fitted with a mechanism probably

---

* During their early colonization of Colombia, the Spanish had learned of a ritual whereby the newly elected leader of the local Muisca people would cover himself in gold dust (the original *El Dorado*, literally, 'The Golden One'), take a raft filled with gold and emeralds into the middle of a lake, then throw them in as offerings. This story was the impetus for a Spanish expedition mounted in 1536, which would ultimately result in the discovery of the local emerald mines. The lake has been identified as Lake Guatavita, and multiple efforts between the sixteenth and nineteenth centuries were made to drain it to recover its secret treasures, none with much success.

commissioned specifically for its crystal casing. But what we know about the Cheapside watch is dwarfed by the many secrets it will never give up. Many more pages are missing from its story. We are none the wiser about the identities of those who respectively extracted such a miraculous crystal, transported it across continents, and cut and fashioned it – a remarkable feat of skill at a time when the relevant techniques would not have been well known.

More mysterious still is the question of where that long and fragile supply chain was due to end. Who was the ultimate owner of such a remarkable and hugely valuable object? Did they commission the watch for themselves, or was it a jeweller's notion, to be pressed on a biddable client? And more intriguingly still, given the nature of its discovery, what of the patron who we assume must first have owned such a stunning piece? That it was discovered among a jeweller's stash invites questions about the status of the owner or intended recipient, and the nature of the transaction. Maybe it was waiting to be collected, had been returned for repair or safekeeping, or perhaps even been put up as collateral by someone who had run into financial difficulties. This masterpiece of jewellery making, with its combination of extraordinary material and visionary craftmanship, might just have been reluctantly surrendered by an owner who had once known and prized it. Or it could never have been collected at all. That we can never know the answer is part of this magical object's charm. Even unearthed from its hiding place, it retains its mystique, holding on to secrets it will never share and a story whose full truth can never be told.

THE SURVIVAL of the Cheapside watch is a miracle that is the sum of many parts. Numerous alternative fates might have prevented it from ever seeing the light of day. It might have fallen victim to one of the shipwrecks that bedevilled merchant transports from the New World. It could easily have shattered under the tools of a craftsman who would probably never have worked on a crystal of this material or magnitude. And, had it not been subject to what was likely a hasty hideaway under the streets of East London, it would almost certainly have suffered the indignity of being cut down and divided into many smaller, more manageable pieces at some point in its history.

Even this inadvertent act of preservation might not have saved the watch, if not for the unusual role played by a curious figure, the antiques dealer and museum curator G.F. Lawrence. During a period of major redevelopment in the capital, it was common knowledge among London's labourers that they would be well compensated for presenting any promising finds to the man they knew as Stony Jack. Discoveries from the magnificent to the mundane were duly brought to him, wrapped in handkerchiefs or traded under tavern tables. The best of these was delivered in 1912, the year after Lawrence had taken on a new role as Director of Excavations at the London Museum. A group of builders called on him one Saturday evening. As a sack was overturned and clods of solid earth spilled onto the floor, one of the party declared: 'We've struck a toy shop I think guv-nor!' Lawrence rinsed off the mud in his kitchen, astonished to find a tangle of Tudor chains and other jewelled treasures.[5] Even better, this was just one of multiple deliveries that came to him over the months that followed, knotted cloth parcels opening to reveal an exceptional collection that was beyond any treasure hunter's wildest dreams. Lawrence had hit the jackpot, for himself and his employers at the London Museum, into whose ownership the Hoard passed, to be exhibited in full for the first time a century later.[6]

It was, therefore, an improbable chain of good fortune that brought the emerald in one piece from the Muzo mine in Colombia to Goldsmiths' Row.[7] And it was more extraordinary still that something so precious and valuable should have survived intact for so long, to be seen and enjoyed by a modern museum audience.

But perhaps none of these quirks of fate is as dramatic as the greatest chance happening of all: that such a handsome crystal ever existed in the first place. The formation of emeralds is a freak of nature. Their complex chemistry brings together beryllium, aluminium, silicon and oxygen, which in its pure form produces colourless beryl, otherwise known as goshenite.* It is only thanks to trace quantities of chromium, and sometimes vanadium, that the vivid green of emerald will appear. But beryllium and chromium do not naturally exist in proximity to each other. This means that it takes unusual geological events to unite these disparate elements: in some cases through an intrusive volcanic surge pushing minerals into rocks where they would not normally be, or, even more rarely, when the necessary elements are picked up at different points along their underground journey and finally deposited together in sedimentary shales to

---

* In the beryl family, pure beryllium aluminium silicate is the colourless goshenite, named after the Goshen Mountains in Massachusetts where it was first found; aquamarine is the bluish-green variety named from the Latin for 'sea-water' and coloured by iron; pink morganite, coloured by traces of manganese, is named after the banker J.P. Morgan, who donated many mineral specimens to the American Museum of Natural History; and yellow heliodor, literally a 'gift from the sun' in Greek, is another iron-coloured beryl.

crystallize out into green gold. The latter is how Colombian emeralds form.

These unlikely geological contortions explain why emeralds are found in only a handful of locations around the world, none more important than Colombia, whose propensity to produce perfect hexagonal crystals is historically unmatched. Not long after I had the opportunity to handle one of the finest early examples, the Cheapside watch, I had the equally exhilarating privilege of visiting the place from which it had emerged, the mine of Muzo. The town is located in one of two mining districts that sit in the shadow of the Andes, where emeralds have been sought and fought over for at least 1,000 years.[8] The Eastern Emerald Belt includes ancient mines such as Chivor: originally known as 'Somondoco', or 'God of Green Gems', and already eked out by the Spanish in the 1530s. The Western Emerald Belt contains the legendary mine of Muzo, kept secret from the Spanish a few decades longer, and already in use hundreds of years before the arrival of the European invaders.[9] From a distance the town is picture perfect, nestled in green hills and flanked by mist-shrouded mountains. Closer inspection reveals an earthier reality: centuries after the Spanish invasion, the region was the subject of bloody fighting for control between competing families for decades until the 1990s (the 'Green Wars'), a conflict that has continued to sputter and smoulder in the years since. When we visited, firearms had not long since been banned from the mines.

The journey to get there can be almost as dangerous. From Bogotá, we weaved our way through the mountains, often on barely passable roads that the rains wash away each spring, leaving only mud tracks. At one point, the local police took pity on us and escorted us part of the way. Our driver had a rosary swinging wildly above his dashboard: an object whose importance became more apparent as we realized his disconcerting habit of occasionally nodding off at the wheel.

Inside the mine itself, the conditions are just as challenging: the temperature tops 40°C and the humidity 80 per cent. Oxygen and electricity are piped in above your head as water flows freely by your feet. Ahead, workers are hammering wooden planks and beams into place, shoring up the next section of the tunnel. The surrounding shale is so dark and sticky that you are given clothes for single use, with little hope that, once stained, they can ever be made clean again.

You have entered an underground factory, a hidden hub of heavy industry thrumming several hundred feet below the earth's surface. Pipes buzz over your head, a constant hum accented by the hammering of drills and occasionally drowned out by the noise of dynamite. Having grown up caving in the Cheddar Gorge, I had long associated subterranean excursions with a peace and solitude that could never be matched elsewhere. That illusion of the underground as a place of quiet introspection was comprehensively shattered by my experiences visiting Muzo and other mechanized mines around the world. It is hard, heavy, sometimes heart-breaking work, dragging the world's most amazing treasures from their hidden homes.

Yet when they are found, the results are extraordinary. On our way back from Muzo, we were invited to visit the mine's factory in Bogotá, where a master cutter sat us down in front of a table laden with dozens of 100-carat stones – an astonishingly hefty haul thanks to a recently discovered emerald-bearing pocket. He demonstrated how these stones would be worked on, and in a moment it was like being transported back in time – as if I were watching the crystal that would become the Cheapside watch being made in front of my eyes. The thought of what it must have been like for the emerald-cutter of four centuries earlier to hold one of these untouched treasures in his hand was almost as exciting as holding the finished product itself had been.

Emerald mining is a dance between the heavy tread of technology and the light fingers of the craftsman. Miners are blasting through the barriers of worthless rock that keep the precious contents hidden, but they are also observing with an expert eye: searching for white whispers of calcite on the dark shale that might – but only might – indicate that emeralds have crystallized within. More fragile than diamond, those crystals would be blown apart by explosives, so miners must move cautiously once they believe a deposit is close, swapping industrial methods for manual tools. It is dynamite one minute, delicate and dainty handiwork the next.

In the heat and the gloom, emerald mining takes the form of a treasure hunt: an enduring quest for a gemstone that has fascinated and enthralled the world for not just hundreds but thousands of years – with a role and status stretching back through many of history's foremost civilizations, and an attraction that the passage of millennia has done little to tarnish or erode.

THE CONTINUOUS status of emeralds through ancient and modern history can be glimpsed through a documentary record that is scattered and fragmentary, shards of evidence peeking through like the first gleam of a newly discovered gemstone.

The earliest clue can be found in a book written by the vizier Ptah-Hotep, a high-ranking official in Egypt's Fifth Dynasty, considered by some to be the first philosopher to have committed his ideas to papyrus. His book of wisdom or maxims, generally dated to around 2400 BC but which may have been several centuries older, is thought to have been produced as a guide for other junior administrators, or even his own son. It includes the first known reference to emeralds, and perhaps gemstones altogether. 'Be not proud because thou art learned; but discourse with the ignorant man, as with the sage,' he wrote. 'Fair speech is more rare than the emerald that is found by slave-maidens on the pebbles.'[10]

A single stray reference might easily be discounted as an optimistic mistranslation, but for the supporting evidence of an active emerald trade in ancient Egypt. Archaeological studies of mines in Wadi Sikait, a desert settlement between the Nile and Red Sea, have suggested that they were exploited as early as 300 BC, centuries before the same settlement became the epicentre of the Roman Empire's emerald industry.[11] We also know, from the painted portraits attached to mummies from the Fayoum buried in the first to third centuries AD, that emeralds were not just mined by the ancient Egyptians but prized among their social elite, who wore them in necklaces and earrings, quite often paired with pearls. Their most famous fan was supposedly Cleopatra, after whom the workings were later named 'Cleopatra's Mines'. The final ruler of the Ptolemaic dynasty, lover of both Julius Caesar and Mark Antony, her story has long been associated with the emeralds and other gemstones she is believed to have worn, adorned her palaces with, and presented as gifts to visiting dignitaries. In his *Pharsalia*, an epic poem detailing Caesar's Civil War, Lucan described Cleopatra attending a banquet: 'a fortune in her hair and around her neck, weighed down with jewellery'.[12]

Emeralds have also been a topic of interest for some of history's most noted anthropologists and lapidaries – ancient authors writing specifically about gemstones. Almost two and a half millennia after Ptah-Hotep began the emerald's literary record, the Roman natural historian Pliny the Elder* effused about the *smaragdus* in his *Natural History*: 'no colour has a more pleasing appearance'.[13] While scattered references to a green stone could be something other than emerald (and he likely includes other green gems under the *smaragdus* umbrella), in some passages it is clearly emerald that Pliny is discussing. At one point he identifies *smaragdus* as a type of *beryllus*, and comes frustratingly close to gemmological accuracy in saying that emeralds are 'polished down into a hexagonal shape by skilled craftsmen' without quite recognizing their natural morphology.[14] Much later, the Persian polymath al-Biruni (973–1048), whose work on minerals included a section on emeralds (which were known as *zammarud* and *zabarjad* – 'two names for the same thing'), highlighted their prominence in contemporary culture, even as he sought to discredit the popular theory that emeralds could be used to blind serpents. He went on to describe the experiments he had undertaken to disprove the notion, from draping snakes in emerald necklaces to making them crawl over emerald-encrusted floors.[15]

We can also infer the popularity of the emerald in ancient Egypt and Rome from references to the practice of producing

---

* Pliny's *Natural History* – an encyclopaedic tome of thirty-seven books on the natural world – was published posthumously by his nephew, Pliny the Younger. As well as a naturalist, Pliny the Elder was also a naval commander, stationed at Misenum across the Bay of Naples from Pompeii when Mount Vesuvius erupted in AD 79. Just as he was observing the huge pyroclastic cloud which had appeared and was making preparations to set out by sea to observe it, he received a letter begging help from a stranded friend, and duly turned his trip into a rescue mission. His friends were saved, but Pliny, taken ill at the end, never returned.

simulants from glass, as well as other materials. The Roman writer Seneca the Elder mentioned boiling up and dyeing other stones to simulate the fractures in emeralds (a technique recorded in surviving papyri on recipes for various gem treatments),[16] while several ancient 'emeralds' in museum collections have been identified as imitations.*[17] In one collection of Roman jewellery I worked on, a pair of earrings had what had always been assumed to have been emerald surmounts, but the green hexagonal beads were actually glass.[18] These were cheap knock-offs in one sense, but sophisticated jewels in another, and proof of the high value already placed on emeralds 2,000 years ago, if the Romans were prepared to produce fakes.

The popularity of emeralds was not only proven by a mass market for forgeries, but also by the perceived need by those in power to clamp down on their use among the plebeians. In AD 529 the Byzantine Emperor Justinian imposed a law allowing any citizen to have a gold ring, but only the emperor was allowed to wear sapphires, pearls or emeralds.†[19]

Emeralds have long existed in myth and legend, spanning numerous cultures. In medieval Europe, they adapted to the

---

* In this case, it is worth noting that the ancient concept of 'smaragdus' most likely encompassed other green gem materials, meaning that what we would today consider 'imitation' might not have been what the Romans considered to be fake. Man-made simulants like glass, however, as well as other materials shaped to look like natural hexagonal emerald crystals, were clearly intentional copies.

† This law seems to have been put in place in response to, quite literally, unbridled luxury among the masses: 'No one shall hereafter be permitted to decorate the bridles and saddles of his horses, or his own belts with pearls, emeralds, or hyacinths [sapphires], or to insert them therein. We, however, permit them to adorn the bridles and saddles of their horses, and their own belts, with other jewels.'

religious environment. The mythical Prester John (John the Priest), a popular Christian folk hero during the era of the Crusades and for centuries afterwards, was widely depicted as carrying an emerald sceptre to denote his wealth and status – an item tied up with the mythology of the Holy Grail, sometimes believed to have been an emerald cup.[20] While in Colombia, emerald mythology surrounds two mountain peaks overlooking Muzo – Fura and Tena. They are named for the lovers who, according to Muísca legend, were respectively the mother and father of humankind, Adam and Eve figures who were promised eternal youth in return for mutual fidelity. When Fura is tempted into breaking this vow, given away by her sudden apparent ageing, they are stripped of their immortality and Tena kills himself, ripping his own chest open. Every tear that Fura sheds over her husband's lifeless body becomes one of the emeralds that the people of Muzo will seek for centuries to follow. In his anger, the God who created these human archetypes turned them into rock, forever to stand guard over the paradise they once called home, separated by the river that continues to cleave a path between them today.[21]

The history of emeralds also helps to draw the contours of trade and diplomacy between societies, epitomized by the stunning 'Mogul Mughal' Emerald, dated to the seventeenth-century reign of Emperor Aurangzeb. One of the world's largest fine emeralds, the 217.80-carat rectangular tablet was engraved on one side with Shi'a verses, and on the other with poppy and leaf motifs. Its Colombian origin is evidence that the Spanish were not stockpiling their most prized emerald finds but trading them, both in Europe and with the great gemstone collectors of the East: the Ottoman and Mughal emperors and the Safavid dynasty in Iran, all keen connoisseurs of a stone that brought to life the huge significance of the colour green in Islam, one associated with Paradise and with the Prophet himself.*[22] So popular and highly prized were emeralds in Mughal India that it was often wrongly assumed that they must have been mined there. Wider European reaction to the 'New World' emeralds coming out of Colombia was, however, one of suspicion, resulting in these unusual new stones initially being traded out of India as eastern gems, a 'greenwashing' of their origin to reassure more sceptical European consumers.[23]

The craftsmen of the Mughal court made the most of the magnificent hexagons that passed through their hands. They not only engraved and decorated emerald tablets, but other *objets* were produced, including emerald drinking cups. The largest known, dating to the early seventeenth century, stands 15 centimetres high, composed of three separate emerald crystals for its foot, neck and bowl.[24] Like the Cheapside watch, its six-sided rim follows the natural shape of the stone, maximizing the outline of the material, while also showcasing the lapidary's art.

This physical practicality is an innate part of the emerald's appeal that should not be underestimated. The emerald is not just a visually stunning gemstone, suited to jewels or deployed as simple decoration. The combination of its size and workability has also allowed skilled hands to mould it into ambitious pieces with utility as well as beauty: yet another dimension to the adoration of a gemstone that has captured imaginations and decorated legends as far back as written records can take us.

---

* Under Ottoman rule, the wearing of green turbans was a privilege extended only to acknowledged descendants of Muhammad.

THE EMERALD'S cultural and aesthetic legacy comes into sharper focus when we consider how these gems are valued and traded today. One of the first lessons I learned in gemmological theory is that there are three vectors of value for any stone: its beauty, its rarity and its durability. Yet the valuation of gemstones in general, and emeralds in particular, so often arises not from quantifiable factors but intangible human associations.

In 2011 an emerald set in a diamond brooch smashed the world-record auction price for the stone, realizing a whopping $6.5 million. This was not only the world-record price for an emerald at auction, but it also obliterated the lot's estimate of $500,000–$700,000. To hear its unvarnished history, this would appear to be unwarranted. It was beautiful and a very fine emerald, but neither the largest, the oldest nor the most magnificent of its kind to have gone on the market. Nor did it have any notably ancient heritage. It was in fact purchased in 1962, over the counter at a Bulgari store in Rome.[25]

What explains the emerald's subsequent value is the identity of the buyer, the actor Richard Burton, and its recipient, his wife-to-be Elizabeth Taylor. That the couple's story was entwined with Cleopatra's, having met while starring in her biopic, only enhanced the connection. And with it, a stone like many others suddenly becomes one whose history is inextricably linked to two of the most famous and glamorous women ever to have lived, making its status as one of the world's most valuable emeralds dazzlingly clear.*

---

* The emerald brooch was the beginning of Elizabeth Taylor's unbroken love affair with the Italian jeweller, Bulgari: 'the only word she knows in Italian', Burton once claimed. 'I introduced Liz to beer, she introduced me to Bulgari.'

Like any gemstone, there are tangible identifiers of an emerald's quality and value, from the depth of its colouring to the extent of its inclusions – small impurities and fissures that affect the clarity of the gemstone, often glossed in emerald terminology as the stone's *jardin* (the French for 'garden'). These are especially common in emeralds, making wholly transparent examples of the stone a considerable rarity. They are immensely useful from a gemmological perspective, often helpful in identifying where the gem was formed. They have also taught me how much we still have to learn. In the case of emeralds, I was told as a young gemmologist that 'three-phase' inclusions displaying a crystal, gas bubble and liquid inside the gem would exclusively denote an emerald from Colombia; now, these features are also found in stones from more recently discovered deposits in China and Afghanistan.

Yet like all gemstones, valuation is about more than the size, weight and quality of the stone. Emerald is a stone for which the story and history are all important. Emeralds from the legendary Colombian mines still sell at a premium even though examples of equivalent quality are being mined elsewhere today. Emotionally and subconsciously, we ascribe value to the certainty of a gemstone that has persisted through countless eras, been loved by many owners, and demonstrated its quality in many different settings. And of course, like in any part of life, we attach a special, even irrational value to what we have chosen for ourselves.

The first emerald I ever bought was while working at Sotheby's, as a gift to myself to mark the first sale I had sole charge over. It was a 1920s Art Deco ring, with a small but vivid emerald – less than half a carat – amid a three-dimensional buckle motif of diamonds. I loved it, wore it everywhere and refused to sell it on to the many admirers who asked. And then disaster struck, when

I continued wearing it while on a trip to Moscow – an eventful trip that I wasn't always certain I was going to survive, and which my beloved emerald did not. I was focused so much on fur coats, snowdrifts up to my knees, and armoured cars complete with AK-47 toting guards, that I didn't give a second thought to what the fluctuating temperatures inside and outside would do to the emerald on my finger. The expansion and contraction must have worked away at a fissure I hadn't noticed, and the cabin pressure on the flight home was the final straw. When I next glanced at my hand, I realized that only half of my beloved beryl was still in its setting. It was an unforgettable lesson in the fragility of the emerald, compared to the tougher ruby or sapphire and the almost indestructible diamond. That the least physically durable of the top-tier gemstones has been perhaps the most enduringly loved through history is one of those delicious ironies that gives gemmology such an insight into humanity.

Our connection to gemstones may begin with a craving for the antique, affection for legacy items with their own history, and attachment to the ones that speak directly to our hearts. But it also runs deeper, travelling beyond the aesthetic and material, deep into the realms of the emotional and the psychological. A career working in gemstones has taught me that various mythical properties ascribed to them since antiquity continue to carry a strong resonance today. As I have learned, the reasons people love emeralds now are much the same as the reasons they loved them a hundred, five hundred and two thousand years ago.

When teaching gemmology, I always begin classes on the emerald by asking what associations students make with their green colour. Usually the answers cluster around a few themes: plants, growth, nature and fertility. The telling thing about these casual, instinctive responses is how closely they mirror the way emeralds have been used and admired by different societies throughout their history. In their cultural birthplace of ancient Egypt, emeralds – and other green gems – had a strong association with fertility, birth and rebirth – draped around the necks of pregnant women and the thighs of those about to give birth, and buried with the dead to guarantee eternal youth in the afterlife.[26]

Alongside this life-giving significance, emeralds have frequently been regarded through history as an aid to health and a cure for diseases from dysentery, epilepsy and leprosy to otherwise deadly fevers. They have been placed under tongues, worn as amulets, and dangled over stomachs to ward off illnesses or as an antidote to poison and specifically snakebite. In many different manifestations, the emerald has carried the connotation of being purifying, cleansing and life-giving.[27]

Beyond medicinal uses, emeralds have also carried a long association with the idea of green as one of nature's most calming, soothing colours. Theophrastus, the Greek philosopher who composed a treatise on gemstones in the fourth century BC, asserted that emeralds were 'good for the eyes', and that people would carry them around to see better.[28] Pliny the Elder described how the Emperor Nero would use an emerald while watching gladiatorial contests – though whether this was as a lens to view the combat, a mirror, or simply something to gaze at to rest his eyes between bouts is not made clear.[29] That anticipates a practice historically common among jewellers, who would place green bowls of water on their workbenches as a comforting contrast for their eyes after hours of painstaking work. A similar observation by the early twentieth-century American surgeon Dr Harry Sherman led to the creation of operating theatres designed entirely in green – a development that began an enduring link with the colour and its soothing properties in hospitals and healthcare settings.[30]

This might all be dismissed as so much dead history, if the associations did not continue today. In one jewellery job, I worked with a pregnant client who was being bought a baby-to-be present. Almost the first thing she said was that it had to be an emerald, because she knew it would be helpful for the birth. Similarly, one student told me how her grandmother would make all the children in her family drink from a tiny emerald cup, to maintain youth and good health – an almost identical association to that enshrined in medieval myths of the Holy Grail.

The emerald's resonant symbolism has helped to ensure that the finest examples are highly sought after. I vividly remember one of the first occasions when a magnificent Muzo emerald came up for sale at auction directly from the trade in 2011, freshly finished and mounted in a ring. The gem was twelve carats, hexagonal in outline, and clean as a whistle. I was not where I wanted to be, in the sale room talking to clients, but sat in a back room, my post-operative knee propped up on a chair after a snowboarding accident. I soon forgot my pain and frustration as I watched the price for the ring creep up and up on the screen in front of me, to the point where it crossed the threshold of $100,000 per carat, then an auction world record for an emerald, a mark that has since increased threefold.*

---

* In 2015, Christie's sold a superb 10.11-carat Afghani gem, which physically stopped me in my tracks when I saw it on view in Hong Kong. Its colour and clarity were better than any emerald I had ever seen, and I knew then that it would give Colombian stones a run for their money: it sold for double the previous world-record, achieving over $200,000 per carat. Two years later that mark was exceeded again and the top emerald spot returned to Colombia when the Rockefeller Emerald – which became known as the Rockefeller–Winston Emerald as the buyer was the famous jeweller Harry Winston – fetched $305,500 for each of its 18.04 carats.

That valuation pales in comparison to the priciest gemstones, rubies and coloured diamonds, the best examples of which can fetch easily in excess of $1 million per carat. This disparity is easily explained. By any reasonable gemmological standard, the emerald ranks as one of the poorer cousins in the gemstone firmament. Thanks to their turbulent formation and unusual composition, emeralds are often highly included (flawed), unlike their sister stones of sea-blue aquamarine which can be as clear as water and grow to much larger crystals. In addition, their propensity to chip or fracture – as I discovered first-hand in Moscow – creates a larger liability than other expensive gems whose toughness is a better guarantee of longevity. But as Pliny the Elder wrote, 'nothing is more green than the green of emeralds'.[31] The value placed in this primary colour has given the emerald parity as one of the 'Big Four' gemstones, alongside ruby, sapphire and diamond, even though they struggle to match up in pure mineral quality and durability.

The remarkable status of emeralds as one of the most widely appreciated and longest-loved gems of all has persisted through history despite their intrinsic flaws. The chromium and vanadium that give them their incomparable colour are, after all, impurities. And the emeralds that enraptured ancient audiences would have been poor specimens compared to the carat-rich crystals that the mines of Colombia, Zambia and Brazil have produced in recent centuries. But that emeralds are hard to find, tricky to work with, and inherently imperfect as gemstones has only added to their attraction. Almost every civilization in recorded history has valued them not because they are dazzlingly flawless, but because they are complex – often as difficult to understand completely as they are to see through entirely. Perhaps more than any other gemstone, the emerald is a mirror of humanity: flawed, fragile, and all the more captivating for these faults. It is the mother nature of precious stones: humanity's longest-loved and most broadly appreciated gem, one that speaks a universal language transcending barriers of time and culture. We prize the emerald because its existence is miraculous, and its appearance symbolizes so many of the things we long for in life: the promise of growth, the guarantee of health, and the prospect of birth, rebirth and enduring legacy.

2

Ruby

The Leader of Gems

*'When a ruby exceeds five carats, and is
perfect, it is sold for whatever is asked for it.'*

JEAN-BAPTISTE TAVERNIER, 1676

'KEEP YOUR HEAD down. Don't even look out of the window.'

I was in the back of a car heading to the gem mines of Myanmar's northern Mogok region, earth's most famous source of the ruby in its brightest, deepest and richest red form. A legendary location for gem hunters, Mogok had long been one of my dreams to visit, but it was also tinged with difficulty and danger. The Venetian explorer Niccolò de' Conti was the first Westerner to report on this mineral Mecca in the fifteenth century. He had travelled in disguise as a local merchant and now I was similarly being told, in no uncertain terms, to stay out of sight.

At the time I visited, Myanmar was still under an American trade embargo and outsiders were effectively banned from the region I was trying to access. My Burmese friends, fellow gemmologists Hpone Phyo Kan Nyunt and Kyaw Thu,* had helped

---

* Kyaw Thu had one of the rarest gemstones in the world named after him a few years later in 2015. Nearly a decade on, only one sample of kyawthuite is known in the world, discovered in Mogok.

me get around this with special passes and permits, but knew better than to risk the attention of any opportunists we might encounter on the long drive from Mandalay, a seven-hour slog up the winding mountain pass known as the 'road of 999 bends'. They told me that, if the wrong people caught sight of us, with my blonde hair in the back of our four-by-four, I was at risk of being kidnapped.

Most of the journey had passed quietly, not unusual in Burmese company, but towards the end Hpone struck up an uncharacteristic spurt of conversation. He was peppering me with questions, clearly trying to distract me. And then, around yet another corner, I realized why. We had climbed, climbed and climbed, and suddenly here it was, looming above us. An arch carrying a sign that I had longed to see ever since I joined the jewellery trade – our equivalent of the big white letters in the Hollywood Hills.

'Welcome To Ruby Land'.

I was finally here, about to tread what is sacred ground in the gem world, a pilgrim completing their journey. My friend, very thoughtfully, had been drawing my attention away so I would not see the famous sign until it was literally towering over us. I am not ashamed to admit that, as I took it in, I felt tears stinging my eyes.

The view from the mountains we had driven up, down to the valley in which the ruby mines, glinting gold pagodas and red rooftops of Mogok are nestled, is one that has long captivated visitors. 'As I look out upon the tortured granite peaks of Mogok, I endeavour to appreciate the stupendousness of [its] antiquity,' the British Army officer and explorer Colin Metcalfe Enriquez wrote in 1930. 'It is as if the gods had favoured the Winding

Valley with their beautiful jewels – red ruby and blue sapphire – and their loveliest blossoms – from temperate peach to tropical orchid – and as if man, for once in his life, had decided to leave Nature alone.'[1]

This Winding Valley is not just a remarkable vista to admire from above. It is also home to one of the most important gem deposits in the world: the Mogok Stone Tract, 70 square miles of gem-containing rock that lies 7,700 feet above sea level. From this magical ground have emerged the most remarkable red stones anywhere on earth. Rubies were definitely being extracted from it and used as tribute by the eleventh century AD and had quite possibly been commercially mined as early as the sixth century. They were almost certainly discovered even earlier: a favoured legend dating to the first century AD depicts a *naga* – mythological half-snake, half-human – laying three eggs, one of which birthed the king of Bagan, the second an emperor of China, and the third the rubies that would be scattered across the valley.[2]

Like all rubies, the stones mined in Mogok are a simple crystalline aluminium oxide compound known as corundum, coloured red by traces of chromium. But unlike the majority of those found elsewhere, Burmese rubies combine the two most desired traits in the red stone: a depth and richness of hue, and an almost supernatural glow emitted when the chromium fluoresces under ultraviolet light, throwing red rays back into the eye of the observer. As is the case with every gemstone, the quality of a ruby depends on minuscule variations in mineralogy, the tipping of the scale one way or the other making all the difference in how a gem will be perceived and valued. The best rubies have a high chromium content, the source of their vital and revered redness:

too little, and instead you get a less valuable pink sapphire. But they must also lack something: the iron content so common in certain ruby deposits, including those found in Cambodia and Thailand, and many of the African stones.

Where iron is present in any meaningful quantity, fluorescence cannot follow, draining the ruby of its apparently magical ability to glow, and dousing the inner fire that has been central to its mystique through history and across cultures. The rubies of Mogok are effectively iron-free: the outstanding (though by no means only) examples of rubies that appear to contain something improbably vibrant, an internal life all of its own that invites comparison with the great forces of nature that share its colour – the sun, fire and human blood. This fluorescence is the Burmese ruby's greatest magic trick, creating the impression of an even redder colour than the ruby actually possesses. Such light-emitting property is not just for show. It has found important practical uses, including in laser technology: a synthetic ruby laser was used to produce the first accurate measure of the distance between the earth and moon in 1969.[3]

The secret of the Mogok ruby's signature hue and glow lies partly in the rock that bears it – ghost-white marble, which began its life as limestone before its fate was altered by one of earth's great geological events, the tectonic collision of the Indian subcontinent with the rest of Asia approximately 50 million years ago. These two landmasses didn't just nudge together – they rammed at full speed, having closed a gap of 6,400 kilometres that once contained an entire ocean, the Tethys Sea.* Although not entirely understood, it is thought that the first collision was

---

* At one point, India was flying along at a rate of 15 centimetres per year: double the speed of the fastest modern tectonic drift.

followed by a much greater impact 25 million years later, approximating to the time when the rubies and sapphires of Burma seem to have been forming.

What followed was a geological rollercoaster: the limestone on the submerged ocean's floor was first forced downwards, where it metamorphosed under extreme heat and pressure into marble. Then into this spanking-clean stone came an intrusion of liquid granite, draining it of its silica content, the common-as-muck mineral that also prevents rubies from forming (by combining with the aluminium itself, in preference to oxygen). Finally, and most dramatically, the continued movement of the two continents after they had bashed into each other created a gradual squeezing and upward buckling effect that continues today, forming the Himalayas and the 'ruby belt' of white marble that the mountains contain.[4]

That seam of pale host rock stretches across the mountain range, spawning rubies from Afghanistan and Tajikistan in the west to Vietnam in the east. All have produced rubies with similar characteristics, and at times of outstanding quality, just not as frequently or reliably as Burma. Consistency, continuity and longevity are behind the reputation of many historic gemstone origins, and it is Myanmar that has both the longest and deepest association with the ruby, making it the spiritual heartland of the stone.

The two factors that often set Burmese rubies apart – colour and fluorescence – are not just intrinsic to its visual appeal. They also help to explain the belief system that has grown up around the ruby throughout its recorded existence: that it is a gem redolent of blood and fire, a symbol of power and protection, a stone strong enough to fight with and valuable enough to fight over. These are the associations that have made the ruby revered through history, and the most highly prized and commercially valuable coloured gemstone in the world. Its superlative status is summed up by its Sanskrit names, *Ratnanayaka* (King of Precious Gemstones) and *Ratnaraj* (Leader of Gems).

The notion that the ruby contains a visceral power – something akin to lifeblood – is tied up in its earliest mythology. The *Garuda Purana*, one of a body of Hindu texts centring on the deity Vishnu, depicts the downfall of Vala, a demon who had defeated several demigods and appeared invincible. When he is finally destroyed by the gods through trickery, his body parts are scattered over the earth – the bones becoming diamond, the teeth pearls, the eyes sapphire and his blood rubies.[5] This association translated across time and different cultures. The fourteenth-century Arab scientist al-Akfani described the finest ruby as having 'the colour of the fresh seed of pomegranate or a drop of blood (drawn from an artery) on a highly polished silver plate'.[6]

Blood has perennially been one of the ruby's primary associations, closely connected to the belief that it could convey protection from harm. Warriors in Burma took this notion to extremes, cutting themselves and inserting rubies under their skin in the hope of protection in the battle to come.[7] Over time that fundamental link between the ruby and lifeblood has evolved, taking on a medicinal connotation – with the ruby considered a healing gem for blood and heart disorders in Ayurvedic tradition – and ultimately becoming an indelible symbol of passion and love, a representation of the heart. Like many coloured gemstones, the journey of its associations has matured from physical to medical and finally emotional – from being a symbol of blood itself to being a figment of what we hold deepest in our hearts.

So universal is the ruby's connection with life and blood that the metaphor has even been taken up by the gem industry, which uses the term 'pigeon blood' to describe the finest (and most expensive) red gems. Used as a quality designation that is somewhat subjective, and often debated, in the strictest sense it denotes several key attributes.[8] A 'pigeon blood' ruby (from the Burmese '*ko twe*', and probably named for the red irises of some breeds of the bird) should display a pure and evenly distributed red colour (a tinge of pink or purple is allowed), strong saturation (intensity of colour), a minimum of fractures or other distracting inclusions, and of course the high fluorescence that provides the ruby's true magic. For many, a 'pigeon blood' ruby must also be untreated, and specifically unheated: no human intervention other than faceting and polishing is allowed to assist its natural state. Almost exclusively, rubies designated 'pigeon blood' have come from Mogok: most other examples fail in some way to meet the punishing threshold, whether because of iron content that suppresses fluorescence, excessive inclusions, or a lightness of tone and lack of colour saturation that pushes a ruby into the territory of being a pink sapphire.*[9]

If the ruby's colour (Latin *ruber*, red) explains the link to blood and vitality, its quality of fluorescence illuminates a second fundamental association – with fire and the sun. This sense of an otherworldly power was evident in early Hinduism, which associated the ruby's inner glow with divine power and frequently made offerings of the stone to the gods.†[10] The ruby's link to the

---

* In the 1980s, in one of the industry's more bizarre gemmological testing initiatives, gemmologist James Nelson sent off to London Zoo for samples of actual pigeon's blood to see if it bore any similarity to the hue of a ruby (brief answer: it did not). He concluded that the pigeon 'can at last be safely removed from the realm of gemmology and consigned back to ornithology'.

† A similar tradition was still apparent in Mogok where we saw old, unlit cabinets stuffed with gems in the local Buddhist temples.

power of the sun is also consecrated in the classic gem combination of the *navaratna* (Sanskrit, 'the nine gems'), an arrangement
of gems popular across Asia as a symbol of health and good fortune, featuring on rings, necklaces, bracelets and *maang tikka*
(jewels worn on the forehead, especially by brides). In the *navaratna*, the ruby sits at the centre, representing the sun, with
eight other gems clustering around it as metaphorical planets.

These symbolic links with human and planetary life forces
have lent the ruby a long-standing association with power, and
the people who wield it. The fourteenth-century travel writer Sir
John Mandeville described how the people of Java ratified their
rulers by awarding them a prized piece of ruby jewellery: 'when
they choose their king, they take him that ruby to bear in his
hand . . . And that ruby he shall bear always about his neck, for if
he had not that ruby upon him men would not hold him for
king.'[11] The ruby has enjoyed widespread use in this way as a signifier of authority: not only the leader of precious stones, but
the precious stone of leaders. In China's Qing dynasty, formal
headwear including gemstone decorations was introduced for
government officials in the 1720s: while middle-ranking mandarins wore hats decorated with rock crystal, coral and sapphire,
the most senior of nine ranks would wear a ruby, or a stone that
closely resembled one, as a sign of their standing.[12]

Whether marking out mandarins, designating kings or being
embedded into the skin of warriors, the ruby as a symbol of power
has carried a strongly masculine connotation. This is a male jewel
in many ways: one that is not sifted from soft soil but hewn from
solid rock, and whose associations are visceral, redolent of battles
for power and the spilling of blood, from the demon Vala's downfall to its use as a talisman in war. This masculine bent is reinforced
by the fact that no woman – considered a potential source of bad

luck – is allowed down the mines in Mogok: a chance of fate that would unexpectedly work in my favour when I got there. Women are, however, closely involved in the industry. They represented the majority of those selling stones in and around the mining area, and by tradition 'kanasé' women were allowed to sift through the 'tailings' washed out of the mines as detritus to rescue any smaller gems. Most surprisingly, the manager of the largest mine I visited was female, a school teacher now in charge of over 100 miners. But the mineshafts themselves are an exclusively and closely guarded male province. Just as the ruby itself is often a man's gemstone, the source of its most famous examples is, at its heart, still a man's world.

WHEN I made my journey to Mogok I was seeking out more than just rubies. This was a personal pilgrimage as well as a professional excursion. I was venturing not just to one of jewellery's sacred places, but somewhere especially meaningful in my own family's history. Burma was one of the theatres where my grandfather, Thomas Hetherington Henfrey, had served with the British Army during the Second World War. His war had begun in East Africa, in command of a band of Ethiopian soldiers, mounted on mules, which specialized in reconnaissance and rapid ambush attacks.* His unconventional work with 'Henfrey's Scouts' earned him a nickname – Somali Joe – as well as the Military Cross. It also caught the attention of Major General Orde

---

* According to Carel Birkby, in *It's a Long Way to Addis*, Henfrey liked to claim that because the scouts were usually mounted, 'they might be classed as cavalry'.

Wingate, the eccentric and iconoclastic Special Forces operative whose unusual talents were being drafted to try and address the disastrous situation in Burma, where the British had been driven into a full-scale retreat by the Japanese invasion in 1942. Henfrey was recruited to join the 'Chindits', a guerrilla force that Wingate had raised to conduct long-range, airborne operations behind enemy lines to sabotage and disrupt its war machine. In jungle conditions, on punishingly long marches and entirely reliant on air-dropped supplies, the Chindits faced some of the most gruelling fighting of the war, with disease and exhaustion exacting a huge toll in casualties. Henfrey, who was already in his forties and had first seen military service in 1919 during the Third Afghan War, would have been one of the oldest men among them. Yet he survived, and was mentioned in dispatches, most likely for his work during Operation Thursday, one of the major Chindit campaigns in the spring of 1944.

It was rubies that led me to look into my grandfather's story. One striking piece of kit from his Chindit days had made it home, where it had pride of place in my parents' living room. It was a map, but with a difference: printed not on paper but two-sided silk, and in great detail. I had always known it was there, but never paid it much attention until my own trip to Burma approached. With Google being blocked by the military government in Myanmar, I quickly discovered that one of the most accurate maps I could access of northern Myanmar was in fact right there on my family's living-room wall.

It was not until well over a decade later that I would discover the extent to which my grandfather's path had really crossed with mine, meeting across a yawning generational valley. It was during the Covid pandemic, while quarantining at my uncle's house in Devon, that I found it: a chest containing papers that had belonged to my grandfather, which had quite possibly been left untouched for decades. Confined to my room during the days and only venturing out while the family slept at night, I started delving into a hidden treasure trove of new information about a man I had never met and had always wanted to know. The stash of documents was better than I could have hoped for, even had I known of its existence. As I eagerly flicked through the pages of his diary, thumbed chalk-drawn maps and scanned his operational notes, I got the same tingling feeling as when opening an old family safe full of precious gems for valuation: the thrilling but slightly intimidating knowledge that I was handling something precious, priceless and unique, and not uncovered for a very long time.

Seeing his handwriting and imagining the voice of his words on the page was mesmerizing, like watching an old photograph come to life in front of my eyes. I had heard that he had a wicked sense of humour and a twinkle in his eye, but this was as if I was actually meeting him for the first time. With the family sleeping, I spent two consecutive nights poring over every word. As I did, one in particular kept rising from the page: Mogok. With his diaries in my hands, it was becoming clear that I had not just been in the same part of the world where he had fought with the Chindits, but in the exact neighbourhood.

This discovery gave a retrospective piquancy to my time in Burma, and brought one particular part of it flooding back with vivid realization. While in Mogok, I had got into a long conversation with the mine manager, while the men were all underground and we were kicking our heels at the surface. She wanted to know why I had made the journey all this way, when there were so many mineshafts I was forbidden from descending. I was cautious about making too much of my family connection, but her gentle enquiry encouraged me to tell her that my grandfather had served in Burma. The response was typical: a smile, a nod, and nothing further mentioned.

That evening, after a long day at the mines, we were driving back through the town as night began to fall. Suddenly a car pulled out in front of ours, as if to form a roadblock, and the driver told me to get out. Panic jolted through me, the warning of kidnap from a few days earlier fresh in my mind. I was curtly told to get out of the car. As a small group of people gathered, I gratefully realized in the dim light that the faces were friendly ones. Among them, the mine manager stepped forward. I heard her telling me that this was the Mayor of Mogok, who had something he wanted to show me. In no time at all, I was being given a tour of the officers' mess in what had been the British Army barracks, complete with its original Edwardian fireplace.

Back then, I had known this was the kind of place in which my grandfather could conceivably have been. But now, with his diaries in my hands, I knew that he must have been there. Sitting in my uncle's living room in Devon, my mind 6,500 miles away, I had the bizarre but beautiful sensation of my life starting to make sense, pieces I had never even known were missing falling silently into place.

It was not the last time that my grandfather, whom I never had the chance to meet, would make an unexpected appearance in my life, the extent of our overlapping travels across time becoming apparent. Sometimes we get a second chance to understand a moment in our lives, looking back at it through eyes that know more and see differently than they did the first time. I may never return to the officers' mess where my grandfather had been, but in many ways I can see it so much more clearly now, in the mind's eye, than I could when I was actually standing there. Everything I have learned since means the memory of that day is so much more vivid than was the experience itself. The real meaning of Mogok in my life – not just the long-dreamed destination of a gemmologist, but a deeply significant place in my family's history – only became clear to me long after I had left.

My visit to the officers' mess in Mogok was of great interest when I was taken there, but one whose meaning felt far more profound only after I had pieced together the family history years later. This was not the first time I had been given a second chance at a first impression. Sometimes a world-record, even life-changing, gem will pass before you more than once, and sometimes that second impression can be even more powerful than the first.

I N 2006 I set eyes on the finest red stone I have ever seen, and, I am still certain, ever will. Except that when I first picked it up, I was anything but overwhelmed at the sight. The cushion-shaped, 8.62-carat Mogok stone singularly failed to grab me. Set in a heavy gold mount studded with diamonds, it looked like a real man's ring: dark, hefty and strangely lifeless, with little of the inner fire and fluorescence that are the Burmese ruby's hallmark. 'Huh,' I thought. 'I know that's worth a lot of money. I bet it's going to set a record.' I could see that in theory the stone had a remarkable colour and fluorescence. But I couldn't *feel* it. My heart didn't leap. I soon found out that although I had been ambivalent, I had at least been right: it duly sold, at auction, to the legendary jeweller Laurence Graff, for a then-record price of $420,000 per carat.

Eight years later the ruby was back in my hand, again in Geneva, and I could hardly believe my eyes. I felt like I was seeing it for the first time. Remounted as it was into a bright platinum ring, I could finally see the magic of this Mogok marvel, which suddenly seemed alive with fire. The ruby now looked like the one-in-a-million gem it was: so bright, so crystalline, so red, that for a moment I couldn't process anything else. I knew this stone, I recognized the perfect cushion-shaped outline, and more than that, the tell-tale and sudden sharp iridescent reflection in the corner of the gem. It was this patch of 'silk' – a plane of minute intersecting needles which threw shards of pink and blue light out at the slightest movement – that branded this ruby as the same stone, a gemmological fingerprint. But this time, this pure and crystal gem dragged my heart upwards, left my skin prickling, and I could feel the slight moisture in my mouth, like a child dribbling over candy. I recognized it immediately, but also a sudden sense of urgency hit me: 'I will never see another ruby like this again in my life.'

What explained these different reactions to the same stone? How had one ruby left me feeling cold the first time and prickling with heat rash the second? Part of it was that my knowledge of the ruby, the nuances of the quality we term 'pigeon blood', and the market that surrounds it, had increased by leaps and bounds in the intervening years. But equally important was the context. When I had first looked at the Graff Ruby, it was a grey February morning, the sky was cloud-covered and the ruby seemed equally inert. The second time, the combination of the ring's lighter setting and the bright spring day helped to transform the appearance of the gem. Bathed in UV rays, it glowed with the inner life that only a ruby can contain. The rich traces of chromium were casting rays of red light back at me, and I was immediately under their spell. Sitting in the auction room I was silently egging it on to reach the unprecedented million-dollars-per-carat, and was frustrated when it fell agonizingly short. But no matter, the result had still nearly doubled the standing world record at the time, and the ruby was more than ratified by its buyer. Laurence Graff was the owner once more, repurchasing the gem he had remounted and sold to a client in between the two auctions.

The Graff Ruby was not just a lesson for me about the ability of gemstones that were formed millions of years ago to adapt and evolve: an old stone gaining new meaning in the right setting. It was also proof of the extraordinary commercial market for the ruby, a perennially pricey gemstone throughout history, but one whose valuation was now scaling new heights at dizzying speed. When the Graff Ruby went under the hammer in 2006, it fetched $3.6 million. A mere eight years later, it had sold for more than double that price: $8.6 million. The price per carat had soared to $997,000, a near miss with the million-dollar threshold. But we had not long to wait. Within just six months, the Sunrise Ruby, a 25.59-carat monster, and an exception in size, came up for sale. As I sat with one of the underbidders, I watched as it reached an incredible $30 million. It had not only become the most expensive coloured gemstone ever sold at auction, but it had hit the magic million mark, selling for $1.185 million per carat. That was in May 2015, and by the end of the year this record had been broken yet again.

It is important to emphasize that there is nothing new about rubies costing the earth. The gemstone's innate visual attractiveness, the cultural importance of the colour red, and the rarity of stones with any significant carat weight have all combined to make it one of the most expensive and sought-after gemstones, second only to certain coloured diamonds in price. If colour and fluorescence define the appearance of the finest red stones, and blood and fire have provided its primary symbolism, then the final piece of the ruby jigsaw through history has been its elevated valuation. 'The price paid for this stone by the Ancients was very high,' wrote the Victorian jeweller Edwin Streeter, who had conducted a first-hand study of the Mogok mines in 1889 and was also financially invested in them. He quoted the seventeenth-century goldsmith Benvenuto Cellini, who had estimated that a top-quality ruby would command eight times the price of a diamond with the same carat value.[13] Much the same had been observed by the philosopher Theophrastus, a contemporary of Aristotle in fourth-century BC Athens, who wrote of *anthrax* – the most valuable glowing red stone at that time, so if not an actual ruby, at least equivalent to one – that 'it has the highest value' and that even a small one cost forty pieces of gold: enough to buy several houses in ancient Greece.[14]

There is no mystery behind the ruby's upmarket price tag: the gem is as rare as it is magnificent, and larger finds are scarcer still. Jean Baptiste-Tavernier, the French traveller and gem merchant who counted Louis XIV, the 'Sun King', among his clients, offered a timeless assessment: 'When a ruby exceeds 5 carats, and is perfect, it is sold for whatever is asked for it.'[15] Streeter and Robert Gordon, a contemporary who also studied the Mogok Stone Tract, similarly testified to how unusual it was to find a ruby that matched size with quality, Gordon going so far as to say that 'large rubies of perfect colour and flawless are mythical'.[16] Whereas the pink sapphire – effectively the same material, simply further down the saturation spectrum for colour – is sometimes discovered in larger crystals, the ruby, with its higher dose of chromium, almost never is. The replacement of aluminium with chromium in the lattice of the crystal is not only unusual, but also appears to have a destabilizing effect on the structure that induces fractures and prevents rubies from growing to any great magnitude.

At over eight carats and with impeccable quality, the Graff Ruby easily passed Tavernier's test. Yet its intrinsic attractions alone could not explain why it had commanded such a punchy price, especially one whose open market value had doubled within such a short time. Like the Sunrise Ruby after it, this reflected a steep and sudden market shift that was lending a premium to world-class gems of all kinds, a trend exaggerated in the case of the richest and rarest rubies, whose relatively steady growth trajectory suddenly took off in the 2010s. Even records set by stones with a starry provenance, such as Elizabeth Taylor's Van Cleef & Arpels ruby ring, given to her by Richard Burton in her Christmas stocking in 1968, were quickly put into the shade. When Taylor's collection came up for sale in 2011 and her ruby sold for a record half a million dollars per carat, it seemed like an anomaly owing more to its celebrity associations than anything else. Yet hindsight proved it was in fact part of a wider trend: that record was quickly exceeded in the next sale season, and would more than double within the following four years.

Among other factors, this price escalator can be seen as a story of two overlapping supply chains. One has been dwindling and the other surging, but both have converged to create a market in which the only way has been up, as steep and sure as the mountain road to Mogok. The only waning has been from Ruby Land itself, where the identified deposits have been all but exhausted, and have certainly stopped yielding mega gems above Tavernier's threshold.* The lack of new supply at this super-premium end of

---

* Conversely, I saw many large gem-quality sapphires being mined during my visit to Mogok.

the market has given added lustre to Mogok rubies that are made available for sale, as old, special stones have been fetched from the safe by long-term collectors eager to take advantage of the rising tide. But while Mogok may be long past its gem-producing prime, at a wider level the supply of rubies is surging: a major discovery was made in Mozambique in 2009, which, within five years, had become the world's largest ruby mine.[17]

The Mozambique deposit was a startling addition to the ruby universe. For much of its history, this has been a stone as scarce as it is prized. Where sources have been found, they have often provided either a limited supply or an inferior quality of gems. Mozambique, however, broadened the ruby picture as never before, with a range of qualities and appearances so diverse as if to run the gamut of all rubies ever mined. The high volume and generally good quality of these stones has, in a short space of time, had an electrifying effect on the ruby market. It has contributed to a pincer movement of rubies becoming more available as their most rarefied source – Mogok – effectively goes extinct. More than ever before, the ruby is now a toy that jewellers can play with in the mainstream of their collections, helping to broaden its popular appeal, at a time when its most desired gems exist in a tiny and seemingly finite supply. Mozambique has also supplied some outstanding stones, including one which, in 2023, achieved the inconceivable and knocked Mogok off the number-one spot. The enormous 55.22-carat Estrela de Fura was astounding not only for its size, but also its vivid, bright and clear colour. I remarked at the time to trade friends how it visually could have been a Burmese gem, and was not alone in its appreciation: it sold at auction for just shy of $35 million, a new ruby – and coloured-stone – world record.

There could hardly have been a more favourable confluence

of events to drive prices up and up. The growth of luxury consumption in China, where red is the universal symbol of good luck, has also played its part in bolstering an already buoyant market. Perennially a prized and pricey gemstone, the early twenty-first century has seen ruby in the pink as never before.

THE GLOBAL popularity of the ruby, which has pushed its valuation to such lofty peaks, rests in large part on the rich emotional associations that red gemstones have always carried – their likeness to blood and fire symbolizing luck and danger, love and life, money and power. There were long periods of history in which other red gemstones held court in this regard, but by the late nineteenth century ruby was in the box seat. Both garnet and spinel had, for different reasons, lost prestige, while red diamonds are so rare as to be incidental. But ruby was rising. The British annexation of northern Burma in 1886, and the granting of a mining licence to a British syndicate three years later, led to the commercial exploitation of Mogok's mines, hitherto closely regulated. This served to increase both the supply and renown of the world's finest rubies, cementing the stone's position as peerless among red gems.

Supply was not the only force driving the popularity and value of the ruby in the twentieth century, when it also benefited from evolutions in jewellery style and fashion. This was the era of widespread cultural exchange between East and West, especially with India, as its princes honed a taste for European luxury and opened their treasuries – filled with miraculous gemstones – in return. The relationships that flourished between Indian Maharajahs and western jewellers would produce some remarkable pieces, and help forge trends in jewellery-making and appreciation that would prove long lasting.

At the forefront of these developments was Jacques Cartier, youngest of the three brothers who built the famous jewellery house into a global force in the twentieth century. He first travelled to India in 1911, where he discovered an appetite for European tastes, with Indian princes pressing him for pocketwatches.[18] Cartier was tantalized by how the subcontinent brought to life the full glory of coloured gemstones, which he found in abundant number in royal treasuries stuffed full of Golconda diamonds, Colombian emeralds (traded with the diamonds through Goa) and of course Burmese rubies, the collections of which could not be found anywhere else in the world.

Cartier's creative fire had been lit: 'Out there everything is flooded with the wonderful Indian sunlight,' he enthused. 'One does not see as in the English light, he is only conscious that here is a blaze of red, and there of green or yellow. It is all like an impressionist painting.'[19] His journeys back and forth contributed to a growing cross-pollination in jewellery design, with both continents intrigued by the fashions of the other. While the Maharajahs made frequent visits to Paris and London, bringing with them old-fashioned family jewels to be updated into European settings, their original Indian jewels started something of a craze on the continent in return, with engraved stones and the *style hindou* becoming a trend in its own right. This marked a sea change, for not long before Europeans had been struggling with subcontinental pieces which readily mixed together gemstones of different colours. 'To European eyes the gemmed ornaments of the women are wont to seem rather gaudy and tawdry, besides rough in setting. The heterogeneous mixtures of stones, which will place a turquoise, an emerald, and a garnet in close juxtaposition, seem strange after the reserve shown in the West as to jewel combinations,' wrote the journalist Mary Frances Bill-ingham after a visit to India in 1895.[20]

By the 1930s, Cartier had established its 'tutti-frutti' signa-ture (a term only associated with the jewels in the 1970s), a multi-coloured combination of carved ruby, sapphire and emer-ald that would not have existed without this specifically Indian inspiration. The influence of the East was altering the Art Deco landscape, adding explosions of colour to the previously mono-chromatic palette. In turn, western jewellers were educating their clients to enjoy rubies in different ways, as the supply of stones became more prevalent.

As well as drawing creative inspiration from his experiences, Jacques Cartier was building an enviable network of clients among India's ruling class, who increasingly opened the contents of their treasuries to him. One notable patron was the Maharajah of Pati-ala, Bhupinder Singh, who made a dramatic entrance into the jeweller's life after he was summoned to Patiala – in southeast Punjab – to present a pearl that had been exhibited at the recent Delhi Durbar on that 1911 trip. Having arrived at the palace and been kept waiting an age, Cartier suddenly 'saw a Rolls-Royce being driven worryingly fast toward him by a man in cricket whites, one hand nonchalantly on the steering wheel as he leaned out of the car precariously'.*[21] It was his first encounter with a highly significant client with whom he would work for decades to come.

---

* A vignette that captures two of the Maharajah's passions: he captained the first Indian cricket team to tour the UK in 1911, and was an avid Rolls-Royce fan, commandeering a fleet of them to convey him to and from the Savoy Hotel when visiting London.

Of the many spectacular pieces their partnership produced, an outstanding ruby example is the Patiala choker: six strands of ruby beads interspersed with pearls, comprising several hundred stones in total. It was worn by the Maharani Yashoda Devi (the Maharajah's wife), combined with two other huge Cartier ruby necklaces, in a photograph of the Maharajah alongside six of his ten wives in 1931. When one Burmese ruby can feel like a rare enough prize, it is quite something to see so many of them strung together, a taste of quite how rich in gemstone wealth the Indian royal treasuries were.

It is only through a stroke of fortune that we can still enjoy the piece today, as it was almost lost to history after being removed from the Patiala treasury, possibly amid the chaos of Partition in 1947. Like so many jewels of the time, it had not been seen since Indian independence, and was assumed to have been sold off and broken up. It did not emerge until the 2000s, unprovenanced and in the trade, now as a bracelet, its pearls and some of its rubies removed. Although it bore the Cartier signature, the firm's specialists initially struggled to identify it. With the company archives being organized according to jewellery type, no record existed for a bracelet matching this description. It was only when someone was looking at the famous photograph of Bhupinder and his wives, in a book of Maharajahs' jewels, that the choker was spotted and the provenance finally pinned down.[22] It was subsequently restored to its original form, using stock stones from the period, and sold at Christie's in 2019.

The restored choker, once part of an even larger ensemble, gives some indication of quite how extravagant the Maharajah's tastes were and how extensive was his access to important gemstones. (He reputedly 'brought truckloads of jewels and stones' for his European jewellers to work with on his regular trips to

Paris and London.)[23] It symbolizes the masculinity of a gem that is mostly mined by men, its fate then determined and designed by men.* And it illustrates how significant the relationship was between European jewellers and Indian princes, one that not only produced showstopping pieces in this vein, but helped to influence jewellery tastes and trends on a much wider level. It was Cartier's deep exposure to India, its love of coloured stones and the sunlight that enlivens them – unleashing that 'blaze of red' – which helped to determine how they are now enjoyed around the world a century later.

The shifting form of the Patiala choker also hints at another side to these often-elusive, ever-valuable gemstones. Because of their status at the top of the gem tree, rubies have a habit of being reshaped, reset and sometimes even rediscovered, before emerging dramatically in a new light. Just like the choker, or my grandfather's diaries, the next great thing may not be what is about to be pulled out of the ground. It could be the one that is waiting patiently, a ruby long held in a collector's safe, or an old piece of jeweller's stock, its provenance forgotten. As the Graff Ruby taught me, and my family history in Burma reinforced, sometimes the greatest finds have been there all along, waiting for the right moment to be revealed, and for their true colours to shine.

---

* Often to be worn by them too. Another lavish Cartier ruby necklace, created in 1937 for Digvijaysinhji of Nawanagar, was designed for the Maharajah himself to wear.

3
Sapphire

The Gem of Royalty

'Sapphires are worthy of God, kings and counts.'

THE TRAVELS OF SIR JOHN MANDEVILLE,
FOURTEENTH CENTURY

S RI LANKA IS the source of some of the oldest and longest-known sapphires in the world. They have been treasured and traded for more than two thousand years, and formed at least half a billion years before that. Yet the actual origin of many of these brilliant blue stones is still something of a mystery. There have been relatively few discoveries of primary sapphire deposits (the natural locations where corundum crystals, having formed deep in the earth under immense heat and pressure, have been pushed towards the surface). Nevertheless, it has been known for millennia where the results of these formations end up: as individual stones eroded from the mother source and washed away, coming to rest in the corners of riverbeds. Those gem-bearing river sediments – known as secondary, alluvial deposits – continue to be panned and mined today, providing rare glints of precious blue peeking out from piles of worthless detritus, sifted by skilled hands searching relentlessly for crystalline riches among the mud.

The island we know today as Sri Lanka has gone by multiple names throughout its history, many of them enshrining its close association with precious stones and mineral wealth. Prince Vijaya, the first Sinhalese king of a dynasty that ruled the island for centuries following his arrival in 543 BC, named his initial kingdom *Tambapanni* – copper-coloured – in recognition of the dark sands found on the island's north-western shore. By the time of Alexander the Great, it had become known as Tabropane to the Greek speakers who were aware of this outpost along the East-meets-West soon-to-be Silk Route. Marco Polo, the adventurer who journeyed there in the thirteenth century, knew the island as Seilam, the origin of the British name Ceylon, a term still used for the island's sapphires today. It was, for him, 'for its size, one of the finest islands in the world . . . It produces many precious gems, among which are rubies, sapphires, topazes and amethysts'. To the traders from other parts of the world it was known variously in Sanskrit as *Ratna Dweepa* and in Arabic as *Jazirat Kakut*: the Island of Gems.[1]

Dig a little deeper into Sri Lanka's history, like a sapphire hunter sifting through the layers of fact, fiction and mythology, and you find jewellery embedded in all of them. Sri Lanka was supposedly the source of the gems that King Solomon is said to have presented as a gift to the Queen of Sheba. It was also one of the near-deadly destinations for Sinbad the Sailor on his sixth voyage in *One Thousand and One Nights*, when he was shipwrecked – again – with little hope of survival on the island then known as Serendib. In true Sinbad style, he would be carried to safety by a river flowing with gems of all descriptions and 'many precious things', giving him a path through the internal estuaries of the island on a raft he piled high with these rich treasures.[*][2]

---

* It was another, similar Persian fairy tale that led Horace Walpole, the eighteenth-century English writer and Whig politician, to coin the term 'serendipity', from the accidental but incredible good fortune befalling the characters of three princes who visited the island.

For as long as the historical record stretches – into the first millennium BC, with Hellenistic evidence of Sri Lankan sapphires traded across empires as early as the time of Alexander the Great – the island has enjoyed a dazzling reputation for jewels, capturing the imagination of authors and adventurers, sovereigns and scholars. Yet to actually see this millennia-old story in action is to witness a humbler reality. While the products of Sri Lanka's rich mineral underbelly are eye-catching, its gem industry almost blends into the landscape: more artisanal than industrial, the product of a cooperative model and incremental approach that has been designed to ensure its longevity and stay true to its traditions. In the region around Ratnapura in particular – the 'City of Gems', a few hours south-east of the capital Colombo – traditional practices are the norm, from mining and cutting to treating and trading. It is common to see gem-cutters in doorways working at the *hanaporuwa*, a hand-cranked polishing wheel which is operated by pulling a wooden bow back and forth, like an overly eager violinist.* While out in the fields, you could easily walk past a sapphire mine and hardly recognize its relevance – a small structure that from a distance seems more like a roofed well, from which gallons of water will be pumped in the hope that something much richer can be brought to the surface.

Some mines are little more than open pits dug to the level where the gem-rich soil is exposed, to be washed in wide bamboo baskets and its contents panned. In the most basic set-up, no holes are dug at all, and a miner can simply stand in the bend of a flowing river, panning the pebbles from under his feet, hoping

---

* I have tried this a number of times, and although the contraption seems simple, operating it well takes years of practice.

to hit a gemmy sediment. This is a modern continuation of the oldest kind of gem gathering there is: sorting through the dirt by hand to recover stones that have been eroded and washed away from the rock in which they formed, their density compelling them to settle in riverbed corners. I have also seen sapphire seekers mining extant waterways with more complex contraptions, balancing on floating rafts and using poles to dislodge deposits that are then picked up downstream to go into the pan.

Alongside the river, dotted throughout the fields, deeper shafts are sunk a few dozen feet down. These cut through the old alluvial riverbeds, burrowing through soil that has accumulated over hundreds or thousands of years to where water once flowed and carried precious stones on their wet and winding journey downstream. Compared to large-scale gemstone mining, like emeralds extracted from the shales of Colombia, or diamonds blasted out of the Kimberlite pipes of South Africa,* with dynamite and mechanical diggers as par for the course, this is a homespun affair. Down a mud mineshaft, packed with leaves and reinforced with wood, a ladder will drop down to tiny tunnels perhaps thirty feet below. From there, hand-operated ropes and pulleys will bring up parcels of the golden-brown *illam* dirt that contains the promise of something extraordinary.

On one of the many trips I made taking students on a tour of Sri Lanka's gem industry, to see the ultimate Island of Gems in action, I was able to experience river mining myself first-hand. Just downstream from one raft dislodging the deposits from the riverbed, I was invited to join the miners and jump into the river. I soon found myself with bamboo sieve in hand, swilling the

---

* Kimberlite pipes are volcanic eruptions originating in the earth's mantle and are the primary source of extractable diamonds.

water over the edges to uncover the denser material and sifting with my fingers through the leftovers. 'Go to the bottom', I was told by my friend Armil who had brought me to the river, but before I could take this advice, a significant yellow sapphire crystal appeared, to everyone's surprise. It was a honey-coloured barrel-shaped prism that was several centimetres long. Almost lost for words, I lifted it to the sunlight and shouted 'beginner's luck' to the cheers erupting around me. This was not the only occasion on which the serendipity associated with the island would seemingly rub off on me. Over the course of many visits, luck proved to be a skeleton key to my stays in Sri Lanka, unlocking many unexpected and wonderful experiences along the way.

My students were not surprised at the golden shade, at least, having learned that sapphires could come in all colours of the rainbow, and that Sri Lanka could produce all manner of gem materials. It was, however, a chance find: while the sapphires hiding in river bends and buried deep below the surface may be plentiful, they are also randomly dispersed and can take an age to discover, even in areas with known deposits. A 20-carat stone like mine might only turn up alluvially every few weeks at best. It can take miners months or more to chance upon the real prize of a 100-carat sapphire, still a relatively small stone given its density, no larger than a quail's egg.*

With its rough, eroded outline and translucent skin, the stone I had found was a bit like a jelly bean. Yet the latent signs of its

---

* One of the most spectacular sapphires in the world is the Blue Belle of Asia, an enormous and beautiful gem discovered in 1926 in Pelmadulla, only a stone's throw from where we were in the river that day. Fashioned into a 392.52-carat cushion-shape, it set a world record when it sold at auction in 2014 for over $17 million.

original form – an elongated double-pyramid, with pointed terminals and six-sided outline – gave it away. It was immediately recognizable as sapphire just from its shape, despite the beating it had taken over time, rolling through the river. This is one of the wonders of sapphire discovery: the survival of sapphire through this water-worn journey is a mini-miracle in its own right, but also a natural quality assurance. The sapphire's tumbling safari from where it originated to where it ends up is testament to its toughness. Corundum ranks 9 on the Mohs scale of mineral hardness, behind only diamond, meaning it can hardly be scratched. But its durability – not breaking into pieces under physical stress – is almost unrivalled. A more delicate mineral such as emerald could never withstand the same journey. The result is a gemmological test of endurance, a survival of the fittest: stones with cracks and flaws will disintegrate first, leaving the finer, more crystalline material intact at the end of its ordeal.

Given the role of chance in making a discovery, it is no surprise that superstition surrounds the business of sapphire mining, something I learned on my second mine visit in the Ratnapura region. I had climbed barefoot down the bamboo ladder, landed knee deep in water, and begun my scramble through the tight tunnels towards the newly exposed areas underground where sapphire – and less valuable garnet, spinel, peridot and quartz – might be found. Crouched in a corner of the most recently excavated subterranean pocket, I was exploring the operation and talking to the mine manager through a translator, when suddenly I heard a commotion, and saw one of the miners scrambling towards us as fast as the cramped tunnel would allow, shouting in Sinhalese. I froze, prickling with fear at the possibility of a mine collapse or flood, a student having been injured, or the thought that our presence might suddenly have been deemed unwelcome.

My expression must have said as much as I turned towards the translator to find her smiling. 'No, don't worry. He recognizes you. He says he's pleased to see you.' A minute of frenzied, intermediated conversation later and the cause was revealed. Within a week of my last visit, three massive sapphires had been dug up there, one of the mine's most significant recent finds. Since this mine was set up as a Sri Lankan cooperative system, where the landowners, machinery operators and miners all took a direct cut from results, this was something from which the miners themselves would directly profit. Now I was being welcomed as a good-luck charm. At the surface, the miners presented me with wax candles, a symbol of the light in the tunnels. Although the candles melted and warped on my journey home, I have kept hold of them, as valuable to me as the sapphires I had pulled out of the river.

After the initial delight at my warm, although unconventional, welcome, I became suddenly aware of the expectation this association would leave behind. What if they didn't make a new discovery this time? Remarkably, I received a message some weeks later that another notable find had followed that second visit; at this point, I would have friends for life.

I was not the only good-luck charm at the mine. Although the majority of workers are Buddhist, many mines had almost unnoticeable small stand-alone shrines dedicated to the Hindu god Vishnu. It was explained to me that, while he was not their deity, he was still a protector of Sri Lanka, who needed respect for the country's riches to show themselves. The message was clear: when it comes to a business as dependent on luck as sapphire mining, you must appease all possible influences, and do everything in your power to shift fate in your favour.

THE HISTORY of sapphires is shot through with chance discoveries, reinforcing its status as a stone forever entwined with both fate and fortune. The sapphires of Montana, the only American state where they are commercially extracted, were first discovered by prospectors during the Gold Rush of the 1860s, who in their myopic search often deemed the blue pebbles irrelevant. When they were finally examined and identified at Tiffany's in New York by the gemmologist George Frederick Kunz, he declared them 'the finest precious gemstones ever found in the United States'.[3]

A century later, Montana sapphires began enjoying a renaissance, thanks in part to a massive deposit at a site known as Rock Creek, a spot I visited up in the Rockies during the summer-mining months that sounded more like the location of a Spaghetti Western than a sapphire mine. High in the hills and surrounded by firs, my stay there was as beautiful as a film-set, but involved an isolated wood cabin, a temperamental generator which needed coaxing in the middle of the night, and a wooden door riddled with bullet holes: reminders not to play too long outdoors at night among the bears and mountain lions. Here there were unique risks of mining in remote environments, and a reminder of the chance occurrences that could play out either way. In sharp contrast to history, I joined the Montana miners in search of sapphires as their primary quest, while microscopic gold dust became only an ancillary dividend to profits.

Sapphires had been unearthed similarly by chance in New South Wales over a decade earlier, during Australia's first gold speculation in 1851. This time, Australian sapphires were accidentally identified when one prospector sent off red gems for testing in the hope they were rubies; they turned out to be garnets, but tiny sapphires were also hiding in the gravel. Australia would become one of the world's biggest sapphire producers by volume in the years to come.[4]

Yet the most glittering accident in the history of sapphires belongs to Kashmir, producer of the most highly valued examples ever mined. The market premium of these stunners is underpinned by both their quality and rarity: the choicest finds were made during the few decades following the original discovery in the 1880s, and the peak mining years were seemingly over by the late 1930s. While sporadic Kashmir sapphire discoveries have been made since, nothing has rivalled this window of discovery ever again, arguably in the entire world of gems. Even while the mines were extant, they could only be exploited during a fleeting summer season when the snows cleared and the ground unfroze sufficiently to permit digging. Chance and scarcity have been at the heart of the story of these gems ever since they first came to light.

The story of that discovery is as seductive as the deep blue that makes the Kashmir sapphire so highly prized. At some point around 1880 a landslide occurred in the Kudi Valley, in Kashmir's Padar region, near the village of Sumjam, revealing a primary deposit of sapphires at an altitude of approximately 15,000 feet. One version of the story is that a goatherd was walking along when a blue stone fell out of the mountainside into his path. Tom LaTouche, a geologist commissioned to study the deposit in 1887, related an alternative tale that had been told to him: a *shikari* (hunter), looking for something with which to strike a light for his pipe, chanced upon a sapphire, 'and finding that it answered his purpose better than the ordinary fragments of quartz he was in the habit of using, carried it about with him for some time, and eventually sold it to a Laholi trader, by whom it was taken to Simla, where its value was recognised'.[5]

Whatever the true origin, there was no doubting the dazzling quality of the sapphires that had emerged. Naturally occurring blue colours are rare enough, but the blue of the Kashmir sapphire is something extraordinary, enticing for its depth of colour and the paradoxical quality that makes it intense yet somehow gentle, both piercing the gaze and inviting the eye. Its hue is almost a special effect, created not just by the bonding of iron and titanium that causes the natural colour, but by minuscule inclusions of crystalline dust-like particles which deflect and intensify certain wavelengths of the incoming light. This is, remarkably, the same Rayleigh Scattering effect of sunlight in the atmosphere that leads us to perceive the sky as blue. A similar light effect has been found in a few sapphires from the more recent deposits of Madagascar, and even Sri Lanka, but nothing can rival the optical and historical magic surrounding the stones of Kashmir. These sapphires seem to exhibit a special subtle blue glow, an almost indescribable life fighting its way out of the stone through their vivid yet gentle colour. While the particles create additional blue light, they also soften its transmission: a mesmerizing combination of soft blue fire I like to call 'electric velvet'. These are stones you could look at for ever, as magnificent and multi-faceted as the sky they are often held to represent.

It is the sapphire's blue colour that has been responsible for much of its long-standing financial value and cultural significance. Above all, it has created an ongoing association with the heavens and heavenly power. Rulers and bishops in the Middle Ages wore sapphires to assert both their power and their divine right to it. Charlemagne, one of the most powerful emperors who ever lived, was supposedly buried wearing his 'Talisman', an imposing pendant featuring a 190-carat pale blue sapphire identified as Sri Lankan in origin, and one of the largest used in a piece of ancient jewellery. It is thought to have been recovered from his neck after his corpse was exhumed in the year 1000 by Emperor Otto III. In 1804 the Talisman was presented to Empress Josephine, consort of Napoleon, together with a fragment of the arm of Charlemagne himself.*⁶

The association of blue with status and virtue is deeply rooted in culture, art and religious iconography in the West. You will struggle, for example, to find a stained-glass window or painting featuring the Virgin Mary in which she is not dressed at least partly in blue. As the colour of the sky, the hue of the heavens, blue denotes the right of a ruler to hold earthly power by divine right. Its closeness to purple – the colour of kings, thanks to the most valuable pigment in antiquity, dye extracted from the purple murex snail – has also reinforced its royal connections. In the Book of Exodus, when the ceremonial garments of the Jewish high priest (*kohen gadol*) are described in detail, it is stipulated that the tunic be made 'completely of blue cloth'. A drape of the same material was supposedly used to cover the Ark of the Covenant, the relic box containing the stone tablets on which the Ten Commandments had been inscribed. When God appears to Moses to give him the tablets on which the Ten Commandments were written, under his feet were slabs of '*sappheiros*', also interpreted as the material for the tablets themselves.[7] There is a parallel in Persian mythology, in which the earth itself rests on a giant '*sappheiros*', reflecting its blue colour onto the sky.†

---

* Perhaps not the most romantic gift in history.
† Herein lies one of the pitfalls of ancient text interpretation: although our word 'sapphire' derives from the Greek '*sappheiros*' (which in turn has roots in Hebrew and Persian origins), these '*sappheiri*' were simply any 'blue stone' in general. In these cases, the stone being referenced was most likely lapis lazuli: highly prized, mined at the time in ancient Afghanistan, and a material perfect for massive carving.

Yet despite their close association with royalty and divinity, the heavens and the sky, sapphires are not necessarily always blue in colour. Through the magic of chemistry, sapphires can span the whole spectrum depending on the 'colouring elements' that enter their mixture. In a case of art imitating nature, it really is a matter of mixing the right trace elements like on an artist's palette to get the right shades. The bonding of titanium with iron within the corundum crystal structure creates the blue of sapphire. Iron alone can induce yellow, or green, which also tends to have some of the blue elements present. Chromium in sufficient quantities can create the red of ruby, and, in lesser amounts, the near-red of pink sapphires. For orange, a mixture of chromium and iron will do the trick, also producing the super-special and highly sought after 'padparadscha' sapphires, an unusual combination of pastel pink and orange. Named after the Sinhalese for lotus flower, 'pads', as they are more affectionately known, are variably (and somewhat more and less romantically) compared in colour to sunsets and salmon.

What nature has not achieved, man sometimes can, with heat treatment often used to bring out the colour of a sapphire or change it entirely. The right application of heat can bond existing traces of titanium and iron to remove any yellow, turning a green sapphire blue, lighten or darken the tone of a blue stone, or minimize the inclusions inside that detract from its appearance.*

---

* Heat treatment is another signature of the Sri Lankan sapphire industry, and another age-old practice. It can be done industrially in modern temperature- and pressure-controlled furnaces, but I also saw – and tried – simpler set-ups in which the stone would be stowed in a crucible and placed inside a charcoal fire, fanned via a blow pipe. I have met some treaters who will endure eight-hour 'tag-team' shifts to ensure a stone undergoes consistent heating over a period of several days.

In the case of some sapphires, the appearance of inclusions can unusually add value. These are star sapphires, cut with a rounded cabochon surface, over which a six-armed star seems to float. This effect, known as asterism, can only exist if the sapphire contains enough intersecting 'silk' – fine, iridescent, needle-like crystals that reflect the light in a flash of sparkling rays. Another link connecting the stone to associations with the sky.

Those inclusions may hold the key to one of the most important and hotly debated questions about sapphire: where it is from and what value it correspondingly holds. All sapphires are old, but some are strikingly more so than others. In the case of Sri Lankan and Madagascan stones – two islands whose geological similarities can be explained by the fact they were once both jammed up against what is now East Africa – they were both formed over half a billion years ago. In contrast, sapphires from Burma and Kashmir may have been created as recently as 20–25 million years ago.[8] Brilliant modern developments in gemmological research – such as the dating of certain inclusions affected by radioactive decay, like zircon – have made it possible in some cases to assess the age of a sapphire and adduce its likely origin. Yet determining the origin is a fiddly, controversial and endlessly debated subject. It is not uncommon to encounter sapphires with multiple gemmological reports each specifying a different geological origin. These are important distinctions when a Kashmir sapphire can fetch hundreds of thousands of dollars per carat, ten times more than a stone from the still-producing Madagascan mines.

The rarity and undeniable beauty of the Kashmir sapphire makes it one of the most expensive and sought-after coloured gemstones. But sapphires are about so much more than provenance and price-per-carat. These gems with their heavenly hues and royal resonance often have a value beyond the financial, their beauty residing as much in the deep personal meaning they represent as the way they catch and scatter the light. As I would later discover, it did not take the most remarkable examples of the gemstone to create one of the most significant sapphire jewels that the world has yet seen.

QUEEN VICTORIA'S sapphire and diamond coronet is a piece I had long admired from afar before I was lucky enough to see and handle it up close. I had already come to know this royal favourite via the depictions that made it famous: the 1842 portrait by Winterhalter, two years after she had married Albert, in which she still appears almost schoolgirlish, the coronet encircling a bun of brown hair at the back of her head; and the 1874 painting by Henry Graves, a melancholy side-profile in which the ageing Queen sits reading, many years after her husband's death, a hand raised defensively to her face, the coronet almost lost in the lace of her ubiquitous white widow's cap.

Then it caught me by surprise, when I visited Harewood House to bring some pieces for display and was shown around by the late countess. We were walking through the living room discussing her pearl collection when I suddenly saw a photograph on the piano of her wearing the coronet. Because it was not a royal jewel but a personal one, a gift to the young Queen from her beloved husband Albert, it had passed down in private family ownership, subsequently a wedding present from King George V to his only daughter, Princess Mary, when she married the future 5th Earl of Harewood in 1922. My eyes must have been out on stalks because the countess saw me looking. Before I could even ask if she still had it, she paused just for a moment and looked down. 'No. Not any more.'

This was several years before the coronet became newly notorious, when it was purchased in 2016 for $6.5 million by an overseas buyer, before being subjected to an export ban by the British government. A year later, with the support of the Bollinger family, it was bought for the Victoria and Albert Museum and saved for the nation. It was while its permanent display was being prepared that it surprised me for the second time, on what I thought was a routine visit for a cup of tea with the curator. We were walking through one of the galleries when, disarmingly offhand, he asked if I had seen the coronet. I was over the moon when, back in his office, he modestly produced this national treasure from the safe, presented me with a pair of blue latex gloves, and left me to delight in its every detail.

The first thing that strikes you about one of the most famous royal jewels – perhaps the favourite headpiece of the woman who for over sixty-three years was sovereign to nearly a quarter of the people on earth – is how small it is. Even knowing that it is just 11 centimetres in diameter is no preparation to appreciate how diddy and dainty a diadem this is. Whereas the Crown Jewels declaim wealth, power and rule by divine right, this royal jewel is personal: an intimate love gift that Prince Albert had worked closely with the jeweller Joseph Kitching to design, matching a sapphire and diamond brooch he had given Victoria the day before their wedding. Almost every part of the coronet was meaningful to the young Queen, from the design, based on the *rautenkranz* (wreath) that featured in Albert's coat of arms, to the inlaid sapphires, which had likely come from older heirlooms, perhaps from jewels given to her by her aunt and uncle, Queen Adelaide and King William IV.[9]

Delicate as it may seem, this is also a robust and versatile piece, one formed of twenty-three separate sections whose hinges allow it to be flexed into different shapes: the completely closed circlet favoured by Victoria in her youth and in her dotage, or the 1920s bandeau that Princess Mary was later photographed sporting, opened out and worn across the forehead. It is a remarkable piece to hold and move in your hands. Its beautiful mechanisms flow effortlessly, ensuring a flexibility of fashion choices across multiple generations of owners and eras of style. Small but strong, adaptable yet defined, it epitomizes the character of the Queen for whom it was originally made.

Most striking of all are the stones themselves, an unusual jumble of different shapes and sizes. Probably Sri Lankan in origin,* these are not the velvet perfection of Kashmir sapphires. Clearly evident in the centremost shield-shaped gem is a common flaw of the sapphire known as 'zoning', where the iron and titanium bonding has not occurred in a particular part of the crystal, leaving a colourless band running through the blue. An ingenious jeweller's trick has been used to conceal this blemish from a distance, enamelling the setting in blue to make the colour reflect back into the stone. I found these imperfections not so much shocking as endearing. They reinforced that this was a piece whose primary value to its owner was not vested in the gems themselves but the layers of family associations surrounding them.

This was a jewel that retained its relevance right across

---

* While Sri Lanka is the most ancient source and most likely origin for most nineteenth-century sapphires, those from Thailand (Siam) were already known to Europeans in the seventeenth century, and Burmese stones even earlier.

Victoria's long reign, a piece she turned to both as a recently married woman and much later as a mourning widow. Sapphires were a symbolic gift of the passionate love that Victoria and Albert shared, and the sapphire coronet became an abiding symbol of the grief that settled over her in the years after his premature death in 1861. When five years later she finally returned to one of her most prominent duties, the state opening of Parliament, it was the coronet that she chose to don, as if wearing it were the closest thing she could muster to having him at her side.

There is of course a financial market in sapphires, which are weighed, examined and analyzed to ascertain value. Yet like all gemstones, they can be worth so much more than the sum of their carat value and place of origin. The story of the stones, the hands they have passed through and the purpose for which they were fashioned, can be equally important. In any other context, the sapphires that bedeck Victoria's coronet would be ordinary gems of no great significance. Instead, they have become a symbol of one of the great royal love stories and a signature of one of Britain's most famous monarchs, helping to narrate the joys and the tragedies of her extraordinarily long reign. To observe the coronet up close or hold it in your hands is to marvel at how something so small could contain so much personal history and meaning.

The British royal connection with sapphires did not end with Albert and Victoria. The brooch that he had given her the day before they were married, and which she wore pinned onto the front of her wedding dress, was later designated as an heirloom of the Crown, to be kept within the Royal Family. 'Albert's beautiful sapphire brooch', as she described it in her diary, has become one of the most recognizable royal adornments, used both day-to-day and on notable occasions.[10] It was worn by Queen

Alexandra at her coronation alongside Edward VII in 1902, and by Queen Elizabeth II at a state dinner to mark the visit of President Kennedy in 1961, where she paired it with a blue tulle ballgown designed for the occasion by her couturier of choice, Norman Hartnell. That evening she also wore a matching suite, the sapphire and diamond necklace and earrings that her father King George VI had given her as a wedding present in 1947.[11] It would become a staple statement set throughout her reign, the necklace's Victorian design altered with the addition of a pendant and bolstered by a matching bracelet made in the 1960s. Like the woman from whom she inherited the mantle, sapphires were long a sentimental choice for Britain's longest-reigning monarch, resonant with family history and personal significance.

Albert's brooch bears a striking resemblance to the sapphire that has perhaps superseded it as the most famous one of all among the Royal Family's collection: the engagement ring of Diana, Princess of Wales, later presented to Catherine Middleton – who would become Princess of Wales in her footsteps – for the same purpose. Set with an oval Sri Lankan sapphire surrounded by fourteen brilliant-cut diamonds, the ring was designed by the Crown Jeweller Garrard and picked out by Diana from a selection offered to her. Although she had chosen not an existing royal gem but one from the jeweller's own commercial catalogue, there is no mistaking the similarity between the ring and Queen Victoria's brooch – marriage ornaments 140 years apart that could have been made to complement each other. And now the ring has become a royal heirloom in its own right, a piece similarly vested with symbolism and personal significance. As Prince William said when interviewed after his engagement in 2010: 'It was my way of making sure my mother didn't miss out on today and the excitement and the fact that we're going to spend the rest of our lives together.'[12]

From Victoria and Albert to Queen Elizabeth and her beloved father, Prince William and the treasured memory of his mother, there is a royal tradition now stretching back almost two centuries of saying it in sapphire. These brilliant blues have become one of the gifts of choice, the natural hue of royalty taking on a softer and more personal tone that has done much to boost its popularity. In the years that followed Kate and William's engagement, there was a noticeable upsurge in demand for sapphires among jewellery clients who would previously have wanted only emeralds or rubies. In China especially, the blue of sapphire was starting to rival the red of ruby in popularity, linked entirely to the notoriety of one royal ring.

Throughout history the sapphire has been a stone that declaims wealth and power, yet it has also come to assume a significance beyond displays of kingly might and divine prestige. Today it is also a gem that tells a story of love, loss and legacy – one that, as well as projecting power, serves to cast new light on its human face and frailties.

THESE ASSOCIATIONS of loyalty and luck, faith and family, were close to my heart on my last trip to Sri Lanka. This time I was in search of something more than just sapphires, and something which I was desperately hoping to find. I was back in the corner of the island I love most: ironically not the actual epicentre of gem excavations, but a nearby escape, a bolthole beyond the main mining, manufacturing and trading towns, somewhere with a heart and spirit all of its own.

After a week or two of leading student groups, going down mines, practising blow-pipe heat treatment and traditional cutting techniques, visiting jewellers, and touring the jam-packed gem market of Beruwala – where I will often be the only woman, and indeed sole westerner, in sight – I am ready for something different. Time after time I have found myself retreating to the same place, to the resort of Bentota on the island's south-west coast. Here it is almost impossibly beautiful and unbreakably serene: its gardens, beaches and temples feel like a world away from the heat and hustle of the city. I love the simplicity of walking the beach at night, the tide lapping over my feet and washing my footprints away behind me, leaving a paradise perennially untouched.

I try to finish every trip in Bentota, but on my most recent visit I was returning with renewed purpose, not simply to unwind but in search of something specific. The diaries of my grandfather I had discovered two Christmases earlier, in the trunk back in Devon, had revealed something that astonished and intrigued me. I was not the first member of my family to come to this hidden-away place that I considered my own. My grandfather had beaten me to it, by some seven decades. His wartime service in the British Army had taken him to what was then Ceylon, a tropical training ground for the war being waged against Imperial Japan in Burma. When I flicked through his diaries for the first time, the name Bentota had jumped off the page. Although he was primarily stationed further down the coast in Galle and Boossa, during his trips across the country my grandfather had also recorded multiple overnight stopovers in Bentota, where he had stayed at the Government Rest House.

I had never heard of this building but was determined to seek it out, to pick up the trail of a relative I had never met but whose life now suddenly seemed very current and deeply personal to me. I knocked on every door and enlisted the help of every local

Princess Margaret photographed by Cecil Beaton for her nineteenth birthday in 1949 in a white tulle Norman Hartnell dress, a five-strand natural pearl necklace and a cultured pearl bracelet by Mikimoto. Credit: Christie's/Bridgeman Images.

The Muzo Emerald, 12.01 carats, mounted with spessartine garnets and diamonds, courtesy of Sotheby's; the Afghan Emerald, 10.11 carats © 2015 Christie's Images Ltd; the Rockefeller–Winston Emerald, 18.03 carats, mounted in a Winston diamond surround, credit: Harry Winston, Inc.; the Mogul Mughal Emerald Tablet, 217.80 carats © 2001 Christie's Images Ltd; the Cheapside Hoard Emerald, set in the centre with a watch movement *c.* 1600 © Museum of London.

Three necklaces in rubies, pearls and diamonds executed for the Maharani of Patiala in 1931 by Cartier Paris, shown on a wax mannequin on an autochrome plate, Archives Cartier Paris © Cartier; the Graff Ruby, 8.62 carats, mounted by Graff with diamonds in platinum, courtesy of Graff; the Sunrise Ruby, 25.59 carats, with diamond shoulders © 2023 Christie's Images Ltd.

Queen Victoria wearing the sapphire and diamond coronet designed for her by Prince Albert around the chignon at the back of her head, as well as the sapphire and diamond brooch given to her by him as a wedding gift in 1840, just under the sash of the Garter, painted by Franz Winterhalter in 1842, credit: Bridgeman Images.

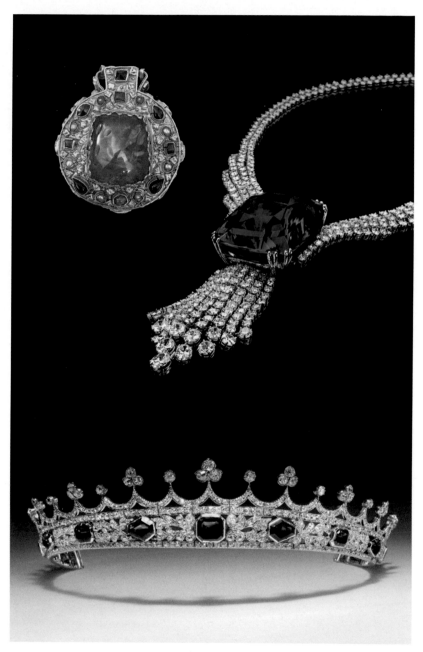

The Talisman of Charlemagne, set with a 190-carat sapphire, in a gold mount with pearls, emeralds and garnets, ninth century © Gérard Panczer; the 392.52-carat Blue Belle of Asia sapphire mounted in a diamond necklace © 2014 Christie's Images Ltd; Queen Victoria's sapphire and diamond coronet, made by Joseph Kitching to Prince Albert's design in 1840 © Victoria and Albert Museum, London.

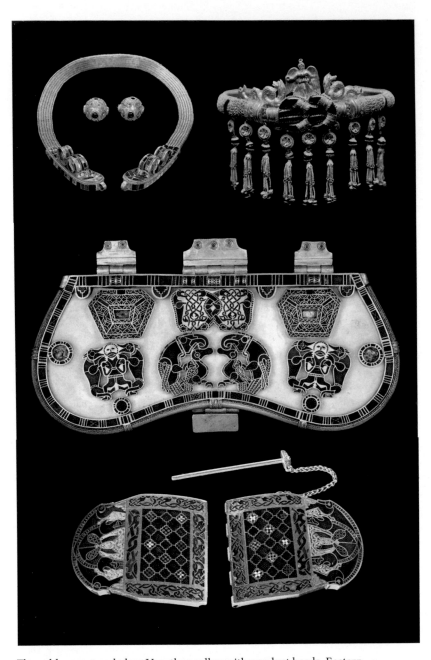

The gold, garnet and glass Umutkor collar, with pendant beads, Eastern Hunnic, fifth century, courtesy of Sotheby's; a Hellenistic gold and garnet diadem, third–second century BC © Vanni Archive/Art Resource, NY; the Sutton Hoo gold, garnet and millefiori blue glass purse lid and pair of shoulder clasps, seventh century © The Trustees of the British Museum.

A tsavorite garnet and diamond 'Papillon' ring designed in 2014 by Glenn Spiro for Beyoncé © Glenn Spiro/Victoria and Albert Museum, London; different garnet varieties in their assorted colours, including a green tsavorite, a yellowish-orange spessartine, a green demantoid, deep and brownish-red spessartines, light and deep pink malaia garnets, and an orange spessartine © yavorskyy.com.

La Peregrina pearl, discovered in the sixteenth century, and owned by Elizabeth Taylor, on an antique diamond surmount © John Bigelow Taylor with permission from Interplanet Productions; Marie Antoinette's pearl, on a diamond bow surmount, eighteenth century, courtesy of Sotheby's; the Armada Portrait of Elizabeth I, representing the defeat of the Spanish Armada in 1588, artist unknown, previously attributed to George Gower, credit: Bridgeman Images.

expert I could think of: the tourist board, tuk-tuk drivers, hotel staff, relatives of people I knew in Colombo, even the local police. With most I drew a blank, but eventually I was pointed to a hotel, a stone's throw from my own very favourite spot, in a place where the coast narrows to a wafer, with the Bentota River on one side and the Indian Ocean on the other. Several conversations with people there confirmed that I had discovered my Holy Grail: it had indeed been built on the site of the old Rest House after it was knocked down. Better still, a few relics had been preserved from those days. I was led into the garden and shown a 150-year-old jacaranda tree, white-flowering and magnificent. A rock pool dating back to the period, overlooking the sea, had also been maintained. The realization washed over me that my grandfather had been here, in this garden, sat nursing a pink gin as the heat of the day subsided, taking in the same cooling view of the ocean that now stretched out before me. He had sought shade under this tree and sat by that pool. The diary even confirmed that he had been the same age then, forty-four, as I was now.

It was a remarkable feeling, as if a portal had been opened allowing me to step back into his life, and he forward into mine. For years I had wondered what all my adventure seeking and gem hunting really represented, coming from an immediate family whose interests had always seemed closer to home. There was no obvious link, no nature or nurture explanation why I am as I am and do as I do. Except there was, and it had been there all this time, hiding in a long-ignored trunk, waiting for me to discover it. As I took my walk along the beach that evening, watching the tide clean up behind me, I realized that I had in fact been following in his vanished footsteps all this time, pursuing a life path whose full meaning I had not known. Some combination of fate, fortune and predestiny had brought us both to this place, joining our paths across the separation of time – just three-quarters of a century, a blink in the eye of a gemstone that may have existed for many millions of years. Like a sapphire being lovingly plucked from the earth, I had recovered something unimaginably precious that time and tide had threatened to bury for ever.

And like the sapphire hunter who strikes lucky, my life had been changed by the discovery. In that line of work, you dig not knowing what you will find, but because the prospect of it is too great to abandon hope. Sometimes, like the long-forgotten trunk being opened or the blue stone falling into the path of an unwitting goatherd, you find something for which you did not know you were looking. More than any other gemstone, sapphires narrate the undulations of life: the low hum of frustration, moments of outrageous fortune, and rare interventions of what can only be described as destiny. The sapphire is so much more than the defining earthly emblem of heavenly blue, significant as that has been through human history. It is also a symbol of the twists of fate that define our lives, the family that connects us, and the moments of life-changing chance that are forever around the next corner – hiding in a bend of the river we cannot yet perceive.

4

Garnet

The Gem of Warriors

*'It is red in colour, and when it is held towards
the sun, it has the colour of a burning coal.'*

THEOPHRASTUS, THIRD CENTURY BC

FOR MOST of us, garnets are red. Small, dark dots that decorate dusty, fusty, old-fashioned brooches. 'When I picture garnets, I think of Victorian housekeepers,' my mother said to me. I knew exactly what she meant. The gems she was calling to mind were Bohemian garnets: clusters of almost opaque, rose-cut stones, nestled together in modest silver or base-metal settings, with perhaps a smooth rounded cabochon in the middle. Close-packed and deep red, these glowing droplets resemble the seeds of a pomegranate, whose Latin name – *granatus* – the gem had adopted.

These Bohemian pyropes were a staple of eighteenth- and nineteenth-century jewellery across Europe.* Although at times worn by tsarinas and queens, Bohemian garnets were not the most expensive of gems, but highly effective decorations in Victorian parures worn on heavy bodices by candlelight. The idea of

---

* An iron-rich garnet variety, named after the Greek for 'fiery eyed': *pyr-ops.*

a slightly dark, fairly cheap, somewhat dated gemstone is garnet's modern reputation in a nutshell.

Yet this does a disservice to a gemstone that has led many lives, all of them well lived. The garnet has actually enjoyed an illustrious, glowing and multi-millennial history, one spent crossing cultures and civilizations, from the ancient Egyptians to modern-day pop stars; and one whose peak, remarkably, reached a technical and artistic age of enlightenment in an era once dismissed as the 'Dark Ages'. Of all the gemstone world's compelling stories, few are more surprising than the rise, fall and rebirth of the garnet.

I WAS mesmerized. Standing in front of these finds, from one of the most famous archaeological discoveries in history, I could not tear myself away. Time had stood still, and then rewound, fast-tracking me back fourteen hundred years until I was there, facing the might and majesty of a great warrior king, through the survival of his jewels. The excavation was Sutton Hoo, and the treasures were the most intricate, artistic and masterful gold and garnet jewels in the world.

In 1939 self-taught archaeologist Basil Brown made the discovery of a lifetime. In a field in Suffolk, under a burial mound assumed to have long since been looted, he struck gold. The Anglo-Saxon longship he excavated would prove to be one of the greatest finds in Britain, and the foremost 'Dark Age' burial in Europe. Inside the wooden hull was a grave filled with riches beyond belief for a burial of this date: luxurious silver plates from Byzantium, bronze hanging bowls with vivid decorative inlays, and a collection of jewels not just fit for a king, but only imaginable in royal ownership.

Although no identification can be certain, that king is thought by many to have been Raedwald, King of East Anglia in the early seventh century, and an early Christian convert among Anglo-Saxon royals.* His massive iron helmet, golden sword pommel, sword belt and cuirass shoulder clasps portrayed a warrior, figurehead, and man of great wealth. These were not only weapons of war, but visual proof (and propaganda) of unrivalled status and power.[†1] And the stone that was chosen to make those statements was the red garnet. The pommel, belt and clasps were all in solid yellow gold, and inlaid with the most extraordinarily detailed geometric dark red garnet designs, accented with bright blue checkerboard millefiori glass.

The craftsmanship is almost inconceivable: hundreds of wafer-thin garnet slices shaped into millimetre-wide motifs, each designed with perfectly angular, and also impossibly rounded, edges, to slot perfectly into an individual gold *cloison*, or cell, of its shape. These jewels were ancient jigsaw puzzles, whose pieces were not crafted from soft wood, but tough and hard gem material, each one ground down by hand, one by one, until it was ready

---

* The Anglo-Saxons had roots in the Germanic migrations of invading tribes of the fifth and sixth centuries, combined with an earlier Roman heritage and cultural overlap with their southern contemporaries in modern-day France (the Merovingians), as well as Scandinavian cultures to the east. Christianity was a further addition to this rich cultural and religious melting pot.

† One unlikely piece of information they yielded about their owner was that due to the wear on the pommel, or handle, of the sword, experts have been able to ascertain that our warrior king was probably left-handed.

to take its place as a miniature piece of a mind-bogglingly intricate design. To ensure each inlay held fast in its cell, a fine gold foil was used behind each slice of garnet to wedge it in place around the edges, but in this even more genius lay: each sheet of gold had been cross-hatched with a pattern intended to show through the fine layer of garnet, so that the effect was like the gold engine-turned patterns behind Fabergé's famous enamels invented fifteen hundred years later; or, as the Curator of Sutton Hoo at the British Museum, Dr Sue Brunning, best explained it, 'a bit like bike reflectors'.[2]

The shoulder clasps, with their additional twisted-rope gold work, and interwound snake-like garnet edging, had me transfixed. Even in the dark museum light, I was clearly in the presence of immense riches, if not royalty. With their rich gold and glowing deep red garnets, the clasps oozed the imperium of the old Roman Empire. I felt something beyond just beauty and value in the exquisite craftsmanship and impressive air of this jewelled armoury: someone extraordinary had worn them, and I could still feel his presence.

There was one jewel from the grave, however, which overshadowed them all: a gold, garnet and glass purse lid. Its singular function was to cover a leather pouch of coins, and it was this find that proved central to the dating of the burial. Inside the purse had been stashed thirty-seven coins, all minted in continental Europe and deposited between AD 610 and 635. This was the world's first bejewelled, blinged-up man-bag. Its gold frame was inlaid with the tiniest garnet slivers, cut into unimaginably complex and minuscule geometric patterns, as well as more visceral, and apparently violent, human and animal hunting scenes. Hounds rise on their hind legs, maws open, pouncing on moustachioed men.* Large birds of prey dig their claws into ducks, whose wing and breast feathers appear to be hand sewn, so finely worked is the detail. Above them are rearing horses, their hooves entangled in a looping design. All these scenes are laid out in hand-worked garnet cloisonné, their cells forming the shapes of a rich story – awesomely beautiful and deeply symbolic. Parallels across Roman, Byzantine, Germanic, Norse and Scandinavian imagery have been suggested for the figures in the scene, together with both pagan and Christian metaphors.[†3]

It was this object that had me rooted to the spot. The longer I looked, the more there was to see: the impossible edges to some of the stones, the appearance of fear or surprise built into the mouths of the men and the wide-eyed ducks, a geometry as well as fluidity in the carving of a crystalline material that is hard to imagine being replicated, even after centuries of evolution in lapidary tools and techniques.

This stunning piece of craftsmanship was part of a much bigger picture. It was in fact the culmination of a millennium in which the garnet had been prized as a gemstone symbolic of power and prestige, and the practice of working with it in jewellery had been elevated to the peak of near perfection. The gold (like the coins) from the burial was Frankish, from the Merovingian kingdom of modern-day France, while the garnets were brought in from as far afield as India and Sri Lanka, having made

---

* I had been so obsessed with the gold and garnet jewels at this point, that I had almost missed that the iconic helmet was also sporting a moustache, albeit much tidier and shorter than the ones worn by the men on the purse.
† It was recorded by Bede that Raedwald had kept both a new Christian and old pagan altar active, as part of his personal transition into Christianity (and as a safe way of hedging his bets).

their way to East Anglia via Mediterranean and Red Sea trade routes pioneered by the ancient Greeks and Romans. The manufacturing itself derived from techniques that lapidaries and goldsmiths had been developing since the international garnet trade became prevalent in the fourth century BC.[4] Craftsmen had gradually honed the art of garnet lapidary work over several centuries, to reach an artistic apogee around the Sutton Hoo collection: shaped edges were ground down by hand with abrasives, while the round eyes in the purse lid's ducks suggest the use of bow-driven wheels, like those used by Roman gem-cutters and which remain a tool of the trade in Sri Lanka today.

The gold and garnet jewels from Sutton Hoo are far more than just exquisite objects to be admired in isolation. In their materials, their manufacturing and their motifs they also help to illuminate an entire period of history, revealing the international sophistication of culture and commerce in early medieval society. The hinges of the purse lid open up an entire world in which gemstones, the techniques for working with them, and the iconography they were used to depict, travelled and intermingled widely. Through understanding the history of the garnet, we may also begin to appreciate the nature of the epoch in which it was so prized – a world whose interconnections were much deeper than we tend to assume, and where the garnet's reputation was so much higher than it would later become.

AT THE centre of this rich mix of gem trading, jewellery making and cultural exchange was the most remarkable gemstone of its age. Long before rubies were discovered in any meaningful number, garnet was the Simply Red of the gemstone pantheon. Carnelian, comparatively orange and opaque, was not as rich, translucent or beautiful in colour; ruby (like spinel) was so rare as to be largely unknown.* For those trading, wearing and writing about red gemstones in early history, garnet was the primary reference point.

To the ancient Greeks and Romans, these stones resembled the 'burnt coals' and 'glowing embers' from which they took their name: the Latin *carbunculus* and Greek *anthrax*. 'It is called anthrax, and seals are cut from it; it is red in colour, and when it is held towards the sun, it has the colour of a burning coal,' wrote Theophrastus, the Greek philosopher who wrote the earliest known treatise on gemstones in the fourth century BC.[5] For Pliny the Elder, writing his *Natural History* in the first century AD, *carbunculus* was the foremost among 'all of the fiery gemstones'.[6] To the early Church Fathers, like Epiphanius, the Bishop of Salamis, in the fourth century AD, this red gem represented more than mere embers. It was like a torch, glowing through one's clothes no matter how hard the attempt to conceal.[7] The supernatural effect of the 'glowing' gems from Sutton Hoo resonated with me, and would not have been lost on its original awestruck audience.

Garnet enjoyed a prestige through the power and vitality, the fire and life force, that red has always represented. It found widespread use both as an adornment to weaponry and in protective jewels including headdresses and hairnets. As early as 3200 BC, garnet was decorating the diadems of the rich, wealthy and elite in Predynastic Egypt.[8] In Hellenistic Greece in the third century

---

* Most translations of valuable red stones in the Old Testament often cite rubies, but in reality the references were almost certainly to garnet.

BC, it was the gem of choice for golden headdresses worn by Greek society ladies. Often combined with the protective amulet of the 'Herakles knot' (evoking how the mythic hero had tied the skin of the Nemean lion around his shoulders) or the talismanic snake (and sometimes both), it was as powerful and protective as it was rich and visual. One of the loveliest examples from this era is a gold hairnet dating to around 200 BC, the intersections of its metallic strings all set with little red garnet cabochons, topped off by the protective image of Medusa.[9] At the same time in Greek-ruled Ptolemaic Egypt, garnets were engraved in cameo with images of the royal family and Egyptian deities, often mounted in gold rings.[10] These usages marked the early flowering of what has been described as the 'Garnet Millennium' – the period from the late fourth century to the seventh century AD in which garnets were the coloured gemstone of choice for the rich, famous and powerful.[11]

While garnet has a history stretching back almost 5,000 years, it really took off in the wake of Greek and Roman imperial expansion. The opening up of trade routes between east and west following the campaigns of Alexander the Great around 330 BC, and later the Roman annexation of Egypt in 30 BC, saw the popularity of eastern luxuries like garnets from India – one of the primary sources of the gem's supply in this period – go through the roof. A sufficiently sophisticated trade had developed between India and Rome that archaeologists have found evidence in south-eastern India of 'large lapidary workshops in the Roman period, producing beads and intaglios for export'.[12] Unlike many other eastern items, the appeal of garnet did not wane with the decline of the Roman Empire, and its cultural relevance continued to grow. One factor was the adoption of garnet by the migratory groups that began to dominate Europe from the late fourth century. Germanic, central Asian and Hunnic tribes became major consumers of garnet jewellery, helping cement its association as a military emblem, as well as popularizing the emerging art of gold and garnet cloisonné techniques.

One extraordinary survival from this period is the Umutkor Collar, a royal eastern piece from the reign of Attila the Hun.[13] The delicately woven gold collar terminates at each end with intimidating garnet cloisonné dragon heads, their crests and eyes blazing with blood-red garnets. The exquisite goldsmithing alongside the inlaid garnets is indicative both of the power of this combination in jewellery form, and of the advanced techniques available to nomadic groups like the Huns. Garnet cloisonné work was as mobile as the tribes themselves, and the production of Hunnic garnet jewellery demonstrates the sophisticated inter-connections of the gem trade of this era: gold sourced as tribute from the Byzantine Empire and the garnets (this time probably from Afghanistan or Pakistan) imported via Persian merchants, already cut and shaped for the purpose. These were the jewels of men, of warriors, of fighters and leaders; and products of advanced workmanship, mature supply chains, and artistic cos-mopolitanism. These were the technical and social precursors to the jewels of Sutton Hoo.

These influences spread west with migration over the fifth and sixth centuries. Several thousand miles away from Attila and his Huns, but only a few years apart, came the burial of the first Merovingian king, Childeric I. His tomb, dating to AD 481, was unearthed in Tournai in Belgium in 1653, bringing to light one of the richest and most significant burials of the era. With him had been interred all manner of treasures, including 300 chunky gold bees, whose wings were inlaid with thick slices of red garnet. Although somewhat stylized, these proto-cloisonné bees would have a lasting effect on history, not only as a precursor to the great artistry of the sixth- and seventh-century Anglo-Saxons and Merovingians, but also for modern-day France. When Napoleon was searching for a suitable imperial motif to decorate his coronation robes as the country's emperor, it was to the garnet bees of Childeric he turned. His adoption of the bee motif simultaneously bypassed the *ancien régime* and connected his reign to the foundation of France.*

Garnet was not only a symbol of military strength and royal status in its era of early renown. Its associations were also spiritual. If trade and geo-politics provided the initial impetus for garnet's popularity, religion would do much of the work to add an extra layer of mystery and meaning. The lapidary baton of Theophrastus and Pliny the Elder was taken up later by Church Fathers including St Ambrose and St Augustine. To these Christian theologians the garnet was more than a symbol of power, prestige and protection; it was the blood of Christ, resurrection and eternal life. Extending the analogy of its blood-like colour, they picked up on the glowing nature of the 'burning coals' or 'fiery embers' of the ancient *anthrax* and *carbunculus*. Ambrose, Bishop of Milan in the late fourth century, made reference in his biblical exegesis *De Paradiso* to 'a bright garnet [*splendidum carbunculum*], in which something of the little flame of our soul lives', indicating the connections that were beginning to be made between the glow of the red stone and ideas of vitality and the spirit. Augustine of Hippo, the North African bishop, noted in

---

* Devastatingly, all but two of Childeric's bees were lost when they were stolen from the Bibliothèque Nationale in 1831 and melted down for scrap gold, but they live on through the imperial emblems of Napoleon, and in the surviving jewels they foreshadowed across early medieval Europe.

his *De Doctrina Christiana* just a few years later that 'understanding the garnet [*carbunculi*], which shines in the dark, illuminates many obscure passages of scripture'.[14] This red gem, gleaming in the darkness, had become a metaphor for both spiritual enlightenment and our very soul.

Christian iconography from later in the Garnet Millennium extended this further. The gem-set pectoral crosses which have been frequently excavated from fifth-, sixth- and seventh-century Christian burials appear in many cases to leverage garnet as an overtly Christian symbol, redolent of the blood of Christ. Over time, this usage of the garnet as a Christian symbol became more theologically complex. According to the eighth-century Benedictine monk Ambrosius Autpertus, the contrasting appearance of the *carbunculus* – apparently dark in colour but then coming alive under light, like the burning coal for which it was named – was itself a metaphor for the incarnation and Christ's two natures, the human and the divine.[15]

Across the centuries of its cultural ascendancy, therefore, garnet was steadily accumulating layers of meaning and significance. For a full millennium, it was the gemstone more than any other that conveyed authority and denoted status. From Hellenistic Greece through to Merovingian France and Anglo-Saxon England, via the Roman and Hunnic Empires, garnet was intrinsic to the elite language of power, decorating objects and jewels used by rulers, military leaders and ecclesiastical figures. It was a trend that crossed centuries, continents, civilizations and world views, finding extensive use in both secular and religious contexts. This was an evolution that reached its peak in the form of the Sutton Hoo jewels, at once masterpieces of a craft that had been developing over centuries, and culminations of symbolism that had travelled widely across cultures.

In its seventh-century examples, the craft of garnet jewellery had reached a dizzying height, after centuries in which the stone itself had enjoyed a spectacular rise. But garnet's travels through time were reaching a roadblock. Much as it rose to prominence with the upsurge in trade brought about by Roman and Greek imperialism, its popularity waned at the end of its Millennium after political upheaval disrupted those same supply routes. This appears to have been the result, at least in part, of the conflict between the Eastern Roman and Sasanian Empires, which were the main mediators of the trade in garnets from India and Sri Lanka via the Red Sea and Persian Gulf. In early medieval Europe, the supply of garnets from the Indian subcontinent was never really replaced. The result was a shift away from the use of Asian to lesser-quality European stones, and an increase in recycling of those from existing pieces.[16] Local sources were found in Portugal, Sweden and Bohemia (the latter only being fully excavated in the eighteenth century), but utilized 'without sustained success, possibly due to the lack of large and transparent garnets'.[17] Garnet jewellery did not disappear, but its status as a dominant feature of the cultural landscape went into abeyance, not to be revived until more than another millennium had passed. These changes marked the beginning of a decline from the dizzying heights of unrivalled garnet gold work; not quite the end of an era, but at least a point after which such artistry, imagination and expertise in garnet jewels could no longer be sustained.

WHEN I first encountered garnets, any concept of great power and politics was beyond me. In fact, my history with garnets began some time before I had ever held one in my hands, and before I even knew what a garnet was. When I was five, my parents decided to install a pond in our back garden. The inevitable summer rains created a slippery pit of sticky, muddy clay, and some of the happiest memories of my child-hood. On the rare occasions my brother and I were allowed to play in this pond-to-be, it was a dream come true. But I didn't just have daydreams about playing in the clay. At night I actually started dreaming of it, through one vision that recurred over and over. I would be digging through the clay heaps – a budding archaeologist even then – and always discover the same thing: a golden cross, its bars widening at the tips, and the entire surface set with flat, polished, deep-red gems, reflecting like mirrors. I would dig up several, put them in a picnic basket under a ging-ham tea towel, and potter down to the village, where my dream would end. I had no idea what these objects were and still don't know what implanted them in my subconscious – perhaps a newspaper or magazine clipping – but the dream was vivid enough that years later when I saw my first garnet cloisonné-work, I recognized it immediately. I had been dreaming of exactly the type of garnet crosses recovered from early medieval Anglo-Saxon and Merovingian graves, employing the same ancient techniques as the treasures from Sutton Hoo.

This strange fragment of my early jewellery obsession would take years to make sense. But in the real world my interest was developing too. When I started visiting local jewellery stores in Somerset with my saved-up pocket money, garnets were all I could afford. The first gem I ever purchased was a garnet, its rich wine-like colour set in silver, a true treasure to my eight-year-old eyes. More gems followed and, for a few years, these pieces were my first jewellery collection. But by my early teens, I was ready to move on. I had learned that garnet was low on the ladder in the hierarchy of red gems, and it was time for me to upgrade. On a family trip to Kenya to visit my uncle, he had agreed to take me to a gem dealer in Nairobi where I planned to add the first ruby to my collection. But when this long-promised visit materialized, with months of carefully saved money burning a hole in my pocket, I was shown only more garnets.

My bitter disappointment at this was, perhaps, an indication of how far the garnet's reputation had fallen since its heyday. The garnets I had collected as a child would have been Bohemian, part of the abundant source from the modern Czech Republic that had dominated garnet supply since the Middle Ages. The Bohemian garnet industry underwent multiple stages of development, culminating in technological and manufacturing improvements which saw production increasingly mechanized by the second half of the nineteenth century, feeding mass-market demand for garnet jewellery.[18] My own adventures in garnet collecting had, unknown to me at the time, come a century after the late Victorian peak of garnet's popularity as a jewel that anyone with a few shillings to rub together could own. This marked an extraordinary culmination of the red garnet's journey through time: from a gemstone that had been the preserve of society's richest and most powerful, a favoured adornment of church leaders and warrior kings, to one that was one of the world's most accessible and widely owned. Aged thirteen, I had little notion that I was turning up my nose at a gem over which the greatest jewellers in the world had once sweated to produce prized items for their most important clients.

Garnets appeared to have fallen from grace, perhaps terminally so. Yet as so often in the history of gemstones, there was a twist in the tale, and it arrived on the back of new discoveries. This development was so unexpected that, to understand it, we have to abandon the whole concept of the garnet as history's red carbuncle. Garnet was about to become so much more than the red gems of royalty, prized by ancient civilizations, then commoditized across the classes. A garnet revival was at hand, and in an entirely different form from its original epoch.

JUST AS the history of garnets is rich and complex, so too is their composition. Most gem families are controlled by a single strict chemical composition, which all their varieties in their many colours will follow. Corundum is aluminium oxide, beryl is a beryllium aluminium silicate, and diamond, quite simply, is carbon.

By contrast, the garnet group is a combination of two series of silicates (silicon oxides), one of which contains aluminium and the other calcium. Among the aluminium-rich garnets are all our ancient red gems, the traditional wine-coloured garnets of the ancient Egyptians, Romans and Anglo-Saxons. Varieties include the deep purplish-red almandine and pyrope – the 'anthrax' and 'carbuncles' of Theophrastus and Pliny the Elder – and the orange-red spessartine, already worked in the Hellenistic period. Rather unromantically, this batch is officially known as the pyralspite series.

The other, calcium-rich class heralds a new chapter for the garnet, containing a surprisingly colourful collection of gems, most of which only became known in the nineteenth century. This group contains the andradite and uvarovite species, with star members including the dramatic deep-green demantoid, first unearthed in Russia in the 1850s. Bridging the gap between the aluminium and calcium silicates are grossular garnets (from the Latin *grossularia*, gooseberry), which contain both aluminium and calcium. Honey-coloured hessonite, also known as 'cinnamon stone' due to its colour and Sri Lankan origins, is a grossular that was discovered early, sometimes appearing in Roman jewels. More recently discovered is tsavorite, a strikingly vivid green East African gem. Collectively, and even less attractively, the calcium-bearing garnets are known as the ugrandite series.

Garnet's mineralogy becomes more confusing still when you acknowledge that these groupings are merely indicative. In reality, garnets often exist on a sliding scale of chemical ratios, meaning a blend of two (or perhaps more) varieties is not unusual. The majority of ancient red garnets are a schizophrenic chemical mix of both almandine and pyrope, a combination so common that the modern trade name 'rhodolite' (meaning 'rose coloured') has become accepted for almandine-infused pyrope garnets.* More recently discovered is a pyrope-spessartine blend, with a little almandine and grossular thrown in for good measure: this unusual blue garnet is made even more curious by its colour-change properties that can make it appear bluish-green under some lighting, and purple in others.[19]

For most of us, garnets were red: not any more.

A T THE same time as the Bohemian factories of the mid-nineteenth century were turning out red garnet for the

---

* Just like the artistic transition between pagan, Roman and Christian iconography in the great garnet treasures of Sutton Hoo, the intrinsic chemistry of these garnets is rarely a clear-cut question. The intermingling of their elements along a sliding chemical scale is a physical parallel to the cultural continuum played out in the motifs they adorn on the jewels.

masses, something dramatic was happening several thousand miles east. Children playing along the banks of the Bobrovka River in the Ural Mountains in western Russia stumbled upon what appeared to be grass-coloured pebbles. Originally assumed to be chrysolite – what we now call peridot – closer inspection revealed that this material was something entirely unknown, distinguished by features not witnessed in any other green gem. Examinations by the Finnish mineralogist Nils von Nordenskiöld* highlighted one particular quality: this stone had the remarkable ability to refract and disperse, scattering white light into its constituent colours like a blazing rainbow. This magic trick is the signature of the colour-less diamond, never before seen in a coloured gem, whose backdrop would normally swallow up competing colours. Nordenskiöld identified it as a type of andradite garnet, coloured by chromium. In 1864, celebrating the 'diamond-like' qualities it exhibited, he proposed the name that stuck: demantoid.[20]

The discovery of demantoid heralded a remarkable about-turn in the fortunes of the garnet. With the red pyrope-almandines having become a bargain-basement gemstone, this new branch of the family would revive associations with the ruling elite. In the hands of some of the world's premier jewellers the demantoid became a feature of the richest contemporary collections, much as the red garnet had been over a thousand years earlier. The stone's brilliance, its ability to disperse light, and its colour – a vivid but deep grassy green – made it both distinctive and

---

* Nordenskiöld had form when it came to discovering Uralian green gems. He was the first to identify as chrysoberyl a stone that emerged in the 1830s with an unusual property: the apparent ability to change its colour such that it appeared green in daylight and red by candlelight (the same mechanism later observed in the blue garnets mentioned above). It was patriotically named alex-andrite, after the then Crown Prince, who would become Tsar Alexander II.

desirable. Its one apparent weakness – asbestos-like curved crystal inclusions known as 'horsetails' – is, unusually, regarded as an essential signature of these Russian stones rather than a regrettable flaw.

The striking appearance and provenance of the demantoid made it an obvious fit for the Tsars of Imperial Russia and their court jeweller of choice, Carl Fabergé. His highly elaborate and detailed work made use of demantoids alongside other Siberian favourites including jade, rock crystal and quartzes. Most notably, the novel green garnet featured in the 'Winter Egg', one of the series that would define the Fabergé name. The Imperial Eggs were created between 1885 and 1917 and presented to the Romanovs at springtime – arriving at Easter as symbols of renewal and rebirth, and helping to commemorate key events in the life and reign of the Imperial family. Crafted in a landmark year, the 300th anniversary of the Romanov dynasty in 1913, the Winter Egg appears frostily pale when closed, its rock-crystal base and oval form decorated with platinum-mounted diamonds and topped with a cabochon moonstone. This lack of bright colour makes the signature surprise hidden within even more striking: a platinum basket of anemone flowers, their white petals crafted from quartz, the pale green leaves in nephrite, and the very centre of each flower, the yellowy-green stigma, represented by a gold-set demantoid. A notably simple design by the highly elaborate standards of the series – and the creation of a female designer, twenty-three-year-old Alma Pihl – the Winter Egg is nevertheless a masterpiece: 'one of the best examples of Fabergé's use of gems to express his composition as opposed to . . . merely as an enrichment', his biographer Kenneth Snowman suggested.[21] The presence of demantoids at the inner core of this piece, a bright peek of colour conveying the promise of spring to come, underscores how its discovery had helped restore status to the garnet – once again decorating the most prestigious jewellery of the world's richest and most powerful collectors.

The discovery of demantoid marked the garnet's return to high status, just as the Winter Egg narrated the hopeful emergence of spring from the long frosts of winter. Yet this was not the end of the stone's revival in new-found form. That story took another turn in the 1960s, in Africa, when a fresh species of green garnet came to light. Every gemstone should have a quirky origin story, and tsavorite's is exceptional. According to the man who uncovered it, Scottish geologist Campbell Bridges, one of his earliest encounters with this career-defining discovery came courtesy of a charging buffalo. This was in 1961, when he was working with the United Kingdom Energy Authority in Zimbabwe (then Rhodesia), seeking out beryllium for use in nuclear reactors. He had taken Sunday off to explore an area of mineral potential suggested by maps and his own observations in nearby hills. As he climbed, suddenly 'an old rogue buffalo charged out of the bush' and Bridges dived for cover into a nearby gully. After shaking off his assailant, he made his way up the hill and, near the summit, was rewarded with the sight of 'small bright green crystals' grinning back at him.[22] Although he didn't yet know it, this was the discovery that would change his life and make his career.

Six years later, he rediscovered these unknown grossular garnets in Tanzania, near the border with Kenya.* He began prospecting and mining in the area, with success, but was forced further

---

* The animal adventures were not completely over. In the valley where this tsavorite discovery was made, Bridges found rhino tracks right by the pit he had dug to explore the deposit. His team had to keep their eye out for the rhinoceros, which was quickly dubbed the 'Mining Inspector'.

afield when his mine was nationalized in 1970. Bridges did what any geologist worth their rock salt would do, and followed the topography – tracing the likely direction of the deposit over the border into Kenya, scrutinizing everything from geological signatures to vegetation patterns and anthills. His diligence was rewarded with the discovery of tsavorite at what would become the Scorpion Mine, which introduced this new green stone to the world.[23]

Tsavorite was an exciting and unusual new find, but it similarly displayed some remarkable characteristics that made it a significant addition to the green-gem universe. It was, Bridges later reflected, 'intrinsically superior to emerald': as well as boasting a cleaner colour that required no treatment, it was less brittle, more brilliant and dispersive of light, and demonstrated greater clarity. He called it the Rolls-Royce of green gemstones.[24] He resisted attempts to name the stone Campbellite after him, instead advocating that it be named for the Tsavo National Park near the land which was its home, and a place which he loved.

Another outstanding discovery was made in Tanzania in 1967 (the same year as Bridges' second tsavorite discovery), not so very far away: a purplish-blue crystalline form of the mineral zoisite. The first gem material found in this family, this new stone showed remarkable properties of its own. Looked at face on, it appeared a beautiful blue, but when the crystal was turned in one direction, it became purplish-pink, and in another, a straw-like yellow.* And like tsavorite, Tiffany's also jumped on it, naming the new gem 'tanzanite'. Advertisements played on the rare and inspiring origin: 'Tanzanite can now be found in significant

---

* It was found that heat-treating the stone removes the less desirable yellow colours, resulting in the pure purplish-blue seen mostly in the market today.

quantities in only two places in the world: In Tanzania. And in Tiffany's.'

These two gems – the deep purplish-blue tanzanite and the vivid bright-green tsavorite – stand as the most important new gemstone discoveries of the last century. Their early history illustrates both the excitement and the challenge of newcomer stones. Both have been picked up by celebrities, such as Beyoncé: the singer appeared wearing an impressive tanzanite ring after the birth of her daughter, and her husband Jay-Z gifted her a stunning tsavorite butterfly ring – now in the Victoria and Albert Museum in London – whose wings actually flutter when worn. The novelty factor is powerful, but so is the challenge of launching something unfamiliar into the jewellery market, even when the mineral in question is of unimpeachable quality. Tsavorite may be preferable in many ways to emerald, but it is also – Bridges estimated – around 1,000 times rarer.[25] When Henry Platt, then head of Tiffany's, asked Campbell how many tsavorites of more than three carats he could guarantee to supply a month, the answer came back that the real question was how many he could get a year. Platt was not deterred: he had already played a major role in the marketing and promotion of tanzanite, and he launched tsavorite in *The New York Times* in October 1974, with a full-page advert declaring it 'an incredibly brilliant green gemstone that is far more durable and far less expensive than emeralds'.[26]

Yet despite the combination of Bridges' outstanding geological sleuthing, and Platt's enthusiastic promotion, tsavorite has continued to lag behind the valuation of top-quality emerald, even though it is a superior gemstone in almost every measurable facet. It's a reminder that, whereas novelty, quality and scarcity can generate value, the gem market is also one that has existed for thousands of years, with buyers whose instincts can be almost as deep-rooted. No new stone, however miraculous, is quickly going to take the place of one that has been mined, traded and collected for centuries. The Colombian emerald – like the Burmese ruby, the Kashmir sapphire or the Golconda diamond – carries a long-earned cachet whose prestige alone adds zeros to the price tag. By contrast, even as a known quantity for over half a century, tsavorite remains comparatively green around the gills as a gemstone. Its rarity – only a handful of other gem-quality sources have been discovered since the initial excavations – is part of its charm, but it is also a further barrier to any aspiration to unseat the emerald from its throne.

A devastating event in 2009 brought the name of tsavorite to the world for all the wrong reasons, when Campbell Bridges was hunted down in the grounds of his property and brutally murdered. For several years, competitors had been illegally prospecting for gems on his land, and issuing death threats demanding that he pack up his operations and leave. In August that year the truck in which he was travelling with a group including his son, Bruce, was ambushed by an armed gang, and Campbell was fatally stabbed. Bruce, who was injured in the attack while trying to protect his father, continues to manage the business and advance his father's legacy, 'carrying on my father's dream for tsavorite'.*[27]

Campbell Bridges's legacy can also be seen in how the garnet

---

* Campbell Bridges's ambition for tsavorite to become internationally admired has advanced considerably since his death, and in no small part because of it: 'If you look at the appreciation of price since my father's tragic murder, tsavorite pricing has shot up exponentially. I would liken it to when a famous painter dies, their artwork goes up in value,' Bruce relates. 'Ironically, my father's dreams became a reality.'

has enjoyed a wider renaissance, as further new varieties have emerged across Africa in the decades since he unearthed tsavorite. Along with Henry Platt, he did not just launch a new gemstone, he helped encourage the trend to give imaginative, emotive names to gems that would once have been given a scientific label guaranteed to set no pulses racing. New discoveries of bright orange spessartine, previously unremarkable, have variously been christened as 'fanta' and 'mandarin' garnets. Certain Tanzanian garnets that do not fit into any of the established groupings – as a hybrid of pyrope-almandine and spessartine that come in shades of orange, pink and red – are labelled 'malaia' (Swahili for prostitute, or outcast). The garnet story has continued to unfold in the region, in a multitude of different colours and chemical forms. Highly unusual blue garnets were first reported in Madagascar in the late 1990s, and some years later in East Africa on the Tanzania–Kenya border.

It is the particular geography of East Africa that has made it one of the world's most important gem-bearing regions. The shifting of the two tectonic plates that underlie the African continent – the Somali plate which encompasses the east coast and islands including Madagascar, and the Nubian plate on which the rest of the landmass rests – has meant good vibrations for gemstones. Millions of years of geological instability along the East African Rift have resulted not just in its complex network of mountain ranges, volcanoes, lakes and rift valleys, but rich deposits of gemstones that snake their way up the coast. Since the second half of the twentieth century, the story of gemstones has increasingly been the story of East Africa, as more of these deposits come to light: from the parallel discoveries of tsavorite and tanzanite in the 1960s, to the unearthing of Madagascan sapphires in the 1990s, Mozambique's rubies in the late 2000s, and garnet in a huge variety of colours and forms.*

As the political landscape of Africa also shifts, local artisanal mining has come to the fore, in an industry now driven by African ownership and entrepreneurship. I had the opportunity to visit one such operation, a small-scale tsavorite mine far out in the bush, started and run by a local Kenyan woman. Miriam took me on a tour of her tsavorite mine not far from Voi, a town on the fringes of the Taru desert. It was one of my more memorable mine visits, in part because I was back exploring the land of my birth, but also for the experience itself. It was a rickety ride into the middle of nowhere to an enclosure surrounded by thornbush fencing that she told me was to protect the giraffes from stumbling into the open pit, but which was also there to protect us from hungry leopards. Miriam's set-up at the mine told a story of pure determination: she ate and slept the business, having set up a shipping crate on site as a bedroom, while surviving on meals of *ugali* (maize porridge) cooked over an open fire, one of which she shared with me and the local Masai *m'zee* (elder). When I visited, it was a small-scale, artisanal mine, a complete contrast to the industrial operations to mine diamonds in South Africa and Botswana, emerald in Zambia, and ruby in Mozambique – much as the tsavorite itself remains the emerald's plucky underdog despite its impeccable mineralogy.

Going there and seeing this nascent green marvel at source – not far from where I was born in Mombasa – helped to complete

---

* Although recently discovered in many cases, the gems of the East African Rift are among the world's oldest, with the formation of many dating back approximately half a billion years.

my journey with the garnet, one which had begun with mysterious childhood dreams and those early pocket-money jewels. Subconsciously I was searching for garnets before I even knew what they were. I would continue to encounter them in unexpected places: like in Pompeii, where as a student archaeologist sorting through layers of pumice I kept encountering dodecahedral black crystals, which I later learned were melanite, a form of andradite garnet.

While I have at various points been dazzled and disappointed by these stones, as a gemmologist I have never failed to be captivated by the story of the garnet and how its fortunes have waxed and waned through history. It is a story of survival and revival, of ancient and modern, of status lost and regained. Stones that were once the boast of royalty became the toast of the cheap and cheerful consumer, before emerging again in new form to grace collections as storied as those of the Romanovs. In its traditional red form, garnet has become the everyman gemstone, while the green branches of its family are among the rarest and most prized of their kind – gems that would be even more valuable if only they were more abundant.

Whether a pyrope that was honed on a lapidary wheel two millennia ago or a tsavorite that was chipped from its host rock within the last half-century, garnets hold a host of information. They also have so much to tell us about the nature of gemstones and our relationship with them as humans who consume, collect, sell, study, search for and fight over these objects. The garnet reveals the full breadth of influences that determine the esteem in which a gemstone is held, and how that is subject to change. Finally, and perhaps most importantly, it reminds us that this reputation is never finally fixed. A stone may have fallen from grace and in value over the course of centuries, but the possibility of its revival is never fully extinguished. Indeed, it may be only a child's chance discovery, a buffalo's charging hooves, or a geologist's determined detective work away from seeing its fortunes rise again.

5
Pearl

The Queen of Gems

'A woman needs ropes and ropes of pearls.'

COCO CHANEL

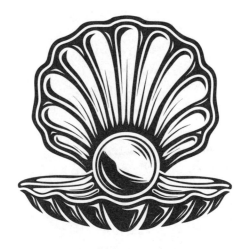

A S SOON AS the box was brought out, I knew that it would contain something special. Unlike those anonymous masterpieces that hide in unmarked containers, this jewel had its entrance announced before it had even shown its face. The dark green leather case, embossed around the edges with gold tooling, bore Princess Margaret's monogram right in the middle: the letter 'M' surmounted with a coronet. As it was pushed across the table towards me, I held my breath, desperate to see what was inside. When I opened the lid – like prising open the shell of an oyster to reveal the pearl within – I let out a sigh of recognition as the necklace appeared. The five strands of natural pearls, on an Art Deco diamond and platinum clasp, was one of the most recognizable, and lovable, jewels in the collection. I immediately felt the warmth of the pearls, an inner glow reflected from the lustrous round surfaces laid out in front of me. The necklace may not have been one of the big diamond showstoppers like the Poltimore Tiara or Queen Mary's diamond rivière, but it was an object of equal significance and perhaps even greater personal meaning.

It was also, although it seems bizarre to say this, a jewel I could completely relate to. The pearls had been a gift to the Princess from her grandmother in 1948, on her eighteenth birthday. Although she herself was royalty, receiving a near-priceless present from Queen Mary, the Queen Mother, the gift was a common British tradition for women: pearls for an eighteenth birthday, or perhaps marriage. Like my mother before me, one of my first 'real' jewels was a pearl necklace (albeit a rather modest strand of small seed pearls), also a present for my eighteenth birthday from my grandmother.

This seemingly unlikely shared experience is one example of the charm, and the success, of pearls across the ages. Synonymous with luxury, romance, power and purity, the pearl is a gem that has represented women throughout history. Grown by chance in the belly of living organisms in rivers and seas, appearing as if by magic with their fully formed, warm, almost glowing surface, and needing no human intervention to improve what nature has already perfected, pearls have become inextricably linked with mystery and a feminine energy. Their iridescent lustre, the way in which light reflects off their skin through countless layers of natural growth, and their innately tactile orb and teardrop forms, make you want to pick them up, touch them, and place them close to the skin.

While ruby may be the Leader of Gems, and diamonds the King, pearl is the Queen of Gems, and the Gem of Queens. It is the perfect term for its role as the most feminine gemstone of all, its history inextricably linked with some of the most famous women in the world. Pearls have been the gem of choice for English debutantes, Hollywood starlets and European princesses, a continuous strand of social history connecting Elizabeth I to Elizabeth Taylor, and Marie Antoinette to Marilyn Monroe.*

Pearls are also a living commentary on the social and cultural changes of femininity and feminism in the twentieth century: from the strict high-necked dog collars of the Edwardian era to the loose, long, swinging single-strand sautoirs of the Roaring Twenties, the wearing of pearls was a micro-record of the emancipation of women in the early twentieth century. Pearls were even one of the Suffragette gems, chosen as a literal message in jewels in the women's movement during the early 1900s. The slogan of Emmeline Pankhurst and her sisterhood, 'Give Women Votes', was spelled out in the colours green, white and violet, and translated into the gems peridot, amethyst and pearl.

By the 1950s, pearls had become the epitome of elegance and glamour, part of the 'New Look' of Christian Dior and now not only worn by the queens of countries and empires, but also queens of the big screen. Pearls immediately make me think of Grace Kelly, Audrey Hepburn, Jackie Kennedy and of course Princess Margaret – all icons of glamour, and not only romantic celebrities, but also powerful leading ladies. These were women who knew what they wanted, and in their fashion choices that was pearls.

A pearl necklace could not have been more perfect for Princess Margaret. A fashion icon, a classic beauty, a British princess and a patron of Dior, like the pearl itself she was feminine yet powerful, both in her style and her character. But perhaps even more poignantly, she was literally named after the gem that she most often

---

* That's not to say they haven't also been hugely appreciated by men, especially in the Far and Middle East, where so many of their natural reserves historically lie: the Indian Maharajahs in particular adored pearls as much as their emeralds, diamonds and spinels.

sported: *Margaret* deriving via French *Marguerite* from the Greek for pearl, μαργαρίτης. Studying and cataloguing her collection, I loved that some of the most important pieces were so personal: the many pearls that spelled out Margaret, and the flower-shaped diamond brooch that stood for her middle name, Rose.*

The importance of this necklace is made apparent by some of the most celebrated images of the Princess. In Cecil Beaton portraits that mark multiple landmark birthdays, the five strands of pearls are repeatedly in evidence. She wears them aged nineteen, photographed in *Vogue*, in a white tulle Norman Hartnell dress embroidered with butterflies and with a subtle rose clutched in her left hand: every inch a princess. The pearls feature again at her twenty-first birthday photoshoot, where Cecil Beaton captured her in an extraordinary off-the-shoulder Dior couture gown, with a seven-layered skirt decorated with golden rhinestones and mother-of-pearl petals. The jewellery accompanying this lavish Cinderella ballgown was devastatingly simple: her wrists and ears were bare, highlighting her favourite pearls around her neck. Their elegant simplicity was all that was needed.

The Princess's pearls say so much about the symbolism, significance and meaning of this gem, and the reason they are such a cornerstone of any jewellery collection. While that necklace features in some of her dressiest images, it was also a consistent favourite in many different contexts. And it was just one part of an extensive pearl collection, including pearls at every level of the value chain. Alongside her precious natural pearls (those

---

* One of my favourite jewels of all was the diamond brooch with the same M-monogram-and-coronet design on the case for the pearls: a motif we appropriated to hallmark each jewel in the collection as proof that they had been part of the official sale.

formed spontaneously in nature), she also had several cultured pearls (also grown in living organisms but by human instigation) and even imitation pearls (cheap but decorative copies of these organic miracles of nature).

For Margaret, pearl was the default gem, the go-to jewel for an independent woman, and something she shared with her sister. Queen Elizabeth II was famous for her own three-strand pearl necklace, a piece so important that she actually had three versions: the original, a gift from her father King George VI, a replica said to have been made to avoid wearing that piece out, and another given to her by the Emir of Qatar in the 1950s. The late Queen was rarely seen without one of these necklaces, from the early period of her reign through to the very last photograph taken of her, at Balmoral in September 2022. It was her uniform and armour, her own gentle fortitude symbolized by pearls. In common with so many women of her generation, it appears that she simply didn't feel fully dressed without them. It is a sentiment best summed up by the legendary designer Coco Chanel. As protests raged through the streets of Paris in 1936, past her boutique on Rue Cambon, she 'decided to go and talk to the rebels', and brushed off advice that she should remove her jewellery. 'Go and fetch my pearls,' she commanded. 'I won't go up to the workshops until I have them around my neck.'[1]

This speaks of the unique status of the pearl, in so many ways a universal fashion statement, accessible to us all. Only the pearl can be equally a special occasion and an everyday piece, an expression of timeless elegance that represents both glamour and simplicity. Pearls can be worn anywhere, on any occasion, suitable for a picnic in the country or going to a ball. They have been some of the most expensive gems ever sold, yet they can also be acquired for the loose change in your purse.

The pearl stands alone also in the gemmological universe: no other gem has been in human hands so long, proven to be so versatile, or become such a ubiquitous ingredient of style. Whereas other gems have seen their reputations fluctuate down the ages, the pearl's reign has been almost uninterrupted. This peerless reputation is well earned. The magic of the pearl does not just derive from its distinctive appearance – the captivating combination of roundness, shape and shine – but from its unique journey into human hands. The pearl is by definition a different beast, as it also comes from a different place.

THE EARLIEST archaeological discoveries of what we would recognize as jewellery date back hundreds of thousands of years. Organic finds at least 140,000 years old paint a picture of our Stone Age ancestors first stringing shell beads in North African caves, in early attempts to signal status and style with the products of the sea.[2] Discoveries that resemble the jewels and gems we prize today are much more recent, and from these it is clear that pearls were probably the earliest precious gems to come into human hands.

Pearls have been recovered from grave sites along the coast of the Persian Gulf, spanning the modern UAE and Oman, that date back to the sixth millennium BC. For more than 7,000 years we have been discovering and prizing pearls, and instilling in them an innate significance: the burials revealed pearls placed above the upper lip and, in later sites, nestled in the hands of the deceased. Some pearls had been strung into bracelets alongside stone beads.[3]

Pearls have been in human circulation for so long simply

because of where and how we find them: not underground but underwater, the product not of vast geological events that date back hundreds of millions of years, but a biological process that happens in less than a decade in the miniature context of a mollusc's humble shell. The pearl is not even a gem*stone* at all, but an organic gem material – what some in the trade now refer to as a biogenic gem, one created by a living organism. As an animal product, pearl is prone to decomposition in a way that its crystalline cousins are not, making those early discoveries all the more remarkable.

To its ancient admirers, the pearl was as rich in symbolism as in material value. The wonder of these pure white gems, born immaculate from the sea, lent itself to vivid imagery. For the ancient Greeks, the pearl was 'the stone of love', and the gem of Aphrodite, goddess of love and beauty: both born of water and risen from the foam of the ocean.[4] According to different traditions, pearls were formed from the tears of the gods or angels falling into the water, or by bolts of lightning hitting the sea.[5]

Most favoured of all was the legend that oysters opened their shells to soak up dewdrops, which became pearls. So extensive was this latter belief that one of the great twentieth-century gemmologists, G.F. Kunz, lamented in his compendious 1908 treatise on the pearl, 'With scarcely a single exception, every recorded theory [of pearl formation] from the first century B.C. to the fifteenth century evidences a belief in dew-formed pearls.'[6] Chief among these was Pliny the Elder, who in his *Natural History* suggested that they emerged from oysters 'filled with dewy pregnancy', the quality of the product depending on whether the sky had been clear or cloudy on the day in question.[7]

That the pearl comes from oceans and rivers explains why it has such a long history – it was discovered by our prehistoric predecessors as soon as they started fishing for food. But it was not formed quite as the ancients believed. From the massive family of molluscs that spans over 85,000 species, only a handful are known to be pearl producing. The majority of these are marine oysters which live in the sea, and freshwater mussels from the rivers, both bivalves, their shells joined in two halves.

The traditional explanation is that pearl formation begins when some kind of foreign body, like a grain of sand, enters the shell and becomes lodged in the fleshy outer mantle, causing the host creature to secrete protective layers of nacre, or mother-of-pearl, around the invading irritant. This nacre is actually a combination of aragonite (a mineralized calcium carbonate, and the same material its shell is made from) and conchiolin (an organic protein), which together are built up in alternating layers, a process sometimes explained as a biological analogue to bricks and mortar.[8] This theory has latterly been challenged by marine biologists who argue that the structure of the shell – through which water regularly passes – makes the irritant idea improbable. They suggest instead that the gem is a reaction to damage to the mollusc's mantle, to which it responds by creating a pearl sac, containing cells that secrete the ingredients of nacre as it heals.[9] So the pearl is not so much the product of an irritation as an injury.

In either case, this process helps to explain the distinctive visual qualities of the pearl: the glow that appears to emanate from within, and the rich rainbow colours that seem to bounce off its surface. Both are a product of the pearl's onion-like layering. As light hits the rounded, finely layered surface of the pearl, some of that light bounces straight off, but some also penetrates into the structure of the pearl, and then reflects off different layers, producing the visual wonder that is lustre – the glow that seems to reach out of the pearl, making it a gem you long to hold.

At the same time, certain light rays reflecting back at different depths hit the edges of the 'bricks and mortar' and are also refracted – split up into their component colours – creating a luminous iridescence of unusual hues. This colour-play caused by diffraction and interference is the same effect seen in rainbow-like reflections off the surface of a soap bubble, and, in gemmology, in multi-coloured opals and the snazzy 'silk' inclusions in ruby and sapphire. Natural variance can augment or detract from these effects: the thicker the layers of nacre, generally the more intense the lustre, whereas thinner layers can produce a milky or even chalky appearance.

The pearl's biological formation helps to explain the qualities of its colour and shape as well as its sheen. Like any other gem, these are determined by the conditions in which it grew, which for the pearl means the host animal that produced it. Pearl-producing molluscs span a variety of species, birthing gems across a wide spectrum of colours, including rare and valuable black pearls (also incorporating shades of grey, green and silver) made by the black-lip pearl oyster, native to the South Pacific, and gems with hues ranging from champagne to bronze, formed inside the golden-lip pearl oyster, found in Indonesia, Australia and the Philippines. Very rare pink or orange pearls might be found in conch or melo shells, both types of gastropods, or sea snails.

Shape varies according to the shell in which it forms. The pearls that emerge can be round, teardrop or pear-shaped, or asymmetrical 'baroques', the latter especially popular in the Renaissance in figural jewellery designs. If the pearl forms attached to the shell, it is called a 'blister pearl' and considered not quite the real deal. Marine oysters consistently beat their freshwater mussel cousins in the lustre and iridescence stakes, and in producing the classically most valuable roundest and whitest treasures.

The pearl is a phenomenon of nature, but no longer a natural creation alone. Once lauded for its miraculous occurrence, the pearl has had its growth environment invaded by the ingenuity of man, as humans have found ways to cultivate its production. As early as the thirteenth century, shaped cultured blister pearls were being produced in China. Freshwater mussels would have small metal Buddha figures inserted into them, inducing the production of nacre.[10] Over the course of centuries the technique has evolved with some success and adapted to different varieties of oysters, but it wasn't until the turn of the twentieth century that a pearl revolution took place, changing the face of the pearling industry for ever.

WHEN PRINCESS Margaret was photographed by Cecil Beaton for her nineteenth-birthday portrait, along with her stunning necklace of natural pearls she wore a bracelet of the cultured variety. It was stamped with the letter M – not for Margaret but for Mikimoto, a pioneer of the cultured pearl. Born in 1858, Kokichi Mikimoto grew up poor, watching his father scrabble to make a living from his noodle shop in Ago Bay, home to Japan's most thriving pearl fishery. Recognizing the enormous European demand for pearls, and that, as a result, the local waters had been heavily overfished, Mikimoto set about finding a way to make the oyster produce the perfect pearl. Settling on the Akoya oyster as the highest-quality host, he spent years experimenting with different types of tissue and bead implants, with terrible tales of massive failures along the way: cracking open endless shells to find that the process had come to nothing, or seeing huge numbers of his crop destroyed by disease. In 1936 he estimated that he had killed over 150 million sea creatures in the course of his work.[11]

He hit the jackpot with his first cultured pearl in 1893, after a devastating case of red tide (a harmful algal bloom) wiped out all but a few of his production, and he earned a patent three years later. By 1905, he had managed to make the pearls round. Mikimoto pearls, as they are still known, were a revelation. A product that could only occur by chance before was now commercially controlled, and it could ultimately be produced in a matter of years rather than decades.

From the 1920s onwards, Mikimoto pearls swept the global market, overcoming the limitations of natural supply to help make the pearl universally accessible, and realize his stated vision to put a pearl necklace around the neck of every woman in the world.* As the art and science of pearl culturing has developed, these gems have emerged in greater number and variety from all over the world: white, silver and gold South Sea pearls, black Tahitian pearls and freshwater cultured pearls that primarily come from China – the latter being the dominant force in the modern industry. The rarity of natural pearls means they are heavily outnumbered today by their cultured counterparts. Pearl may be the Gem of Queens – but Mikimoto was the Pearl King, a man who changed this jewel's story for ever.

---

* The baseball player Joe DiMaggio famously gave a Mikimoto cultured akoya pearl necklace to Marilyn Monroe on their honeymoon in Japan in 1954 – one of her few real jewels, and one which she continued to wear even after their divorce, she said, because it reminded her of happier times.

Unsurprisingly, one immediate (but not permanent) shift was in the downwards direction for natural pearl prices, which, in competition with indistinguishable cultured alternatives, by 1930 had crashed by 85 per cent.[12] They bounced back higher than ever before, largely thanks to developments in pearl testing through radiography. Not unlike a hospital exam, a pearl's organic structure can be explored through X-ray to identify whether there is an artificial nucleus or whether the pearl is entirely a product of nature.[13]

For those not in the price range of natural or cultured pearls, there is a third option: 'pearls' that do not come from nature at all. These imitation pearls share only an external appearance with the real thing, made of an entirely different material (and most often plastic today). They were popular in ancient Rome, where methods for producing them included a technique of foiling glass beads and then adding a top layer of glass. Other approaches at different times have involved coating beads in a bizarre array of substances, including fish scales, the ground teeth of a dugong (a marine mammal known as the sea cow), and even a concoction including egg white and snail slime.[14] While methods have varied, the meaning of these imitations has not: the pearl has been so widely prized and sought after that there has long been a market in producing versions of it outside the complex natural supply chain, and at much more affordable prices. Rather than detracting from the value of the pearl, imitation versions prove its enduring relevance and popularity.

THE WIDESPREAD demand for pearls is rooted in their long-standing symbolic values. The foundational texts of

Christianity, Hinduism and Islam make clear that the pearl's associations with majesty and virtue in particular are ancient ones. In the Quran, pearls hang from the trees in the paradise to which righteous believers are admitted and where they are handed bracelets of gold and pearl. Similarly, the Book of Revelation describes the gates to the kingdom of heaven as 'twelve pearls, each gate made of a single pearl'.*[15] While in the *Rigveda*, the Hindu god Krishna is described as retrieving the first pearl from the ocean and presenting it to his daughter on her wedding day: an early allusion to the pearl's association with bridal purity and virginity.

For the Romans, the pearl was the ultimate symbol of status, and an allusion to the luxuries of – and power over – the Orient. Pliny the Elder noted that the pearl held 'first place among all things of value', and dated the Roman pearl craze to the aftermath of the Third Mithridatic War in 61 BC, where the gem represented the Republic's victory over the Pontic Kingdom and annexation of territories including Syria and Judea. The victory parade of Pompey, architect of the war effort, featured several dozen pearl crowns that had been brought back from the East, as well as a portrait of himself rendered in pearl.[16] Pompey's contemporary, sometime ally and finally enemy Julius Caesar was another of the era's great pearl admirers. 'They say that he was led to invade Britain by the hope of getting pearls, and that in comparing their size he sometimes weighed them with his own hand', recorded the historian Suetonius in his life of Caesar.[17] While it is unlikely he launched his invasions of Britain specifically in search of pearls, Caesar knew that freshwater river pearls existed, and may have been searching for trophies to match those paraded by his

---

* This is the origin of the concept of heaven's 'pearly gates'.

rival, only to be disappointed by what the local mussel population had produced.[18]

Caesar may not have found what he was looking for in Britain, but his appetite for pearls was undimmed. Suetonius also records him gifting his mistress Servilia a pearl valued at six million sesterces (for comparison, the entire fortune of Marcus Licinius Crassus, Caesar's patron and possibly the richest man in ancient Rome, was estimated at 200 million),[19] and how, in 46 BC he sought to impose restrictions on all who could wear the pearl 'except those of a designated position and age, and on set days'.[20]

Already in ancient Rome, the pearl was developing from an elite emblem into an accessible fashion statement. Writing over a century after Caesar's death, Pliny commented that 'nowadays even poor people covet them', complaining that women would wear at least two or three in each ear so that they jangled as they walked. Although he was likely referring more to the Roman middle classes, this indicates – as does the evidence of imitation pearls – that there was a market well beyond the wealthiest echelons.[21]

Yet despite this broader appeal – a hint of the democratization of the pearl that would fully become reality two millennia later – for the Romans the pearl was ultimately a symbol of both military and spending power. That is underlined by the most famous pearl story of the era, and perhaps of all time, a game of one-upmanship between Mark Antony and Cleopatra that was retold by Pliny. Betting who could host the most expensive meal, Cleopatra raised the stakes by declaring that she would spend ten million sesterces on a single dinner. When the final course came, a goblet of wine vinegar was placed in front of her: she detached one of the gigantic pearls from her ears, dissolved it in the liquid and drank it.*[22]

Whatever its veracity, this story has stood the test of time as an illustration of the pearl's association with incomparable value. It also spotlights another of the pearl's most important connections: women and power. From female deities of the Greek and Roman pantheon to some of history's most prominent female monarchs, pearls have been a constant symbol of a woman's ability to rule.

One of the great flagbearers of this tradition was Queen Elizabeth I, for whom pearls were not just attractive accessories, but full-blown forms of propaganda. Famed for her pearls, they carried as many layers of symbolism as their natural nacre. She had built a reputation as the Virgin Queen, chaste and virtuous: characteristics perfectly embodied in the pearl. Although this had long been a part of the pearl's story of symbolism, the association had been reinforced during the Renaissance, when the gem became a favoured feature in depictions of the Virgin Mary and Child.[23] Once an ambiguous religious symbol partially associated with excess and vice – in the Book of Revelation, the whore of Babylon is 'decked with gold and precious stones and pearls' – the pearl was increasingly seized upon as a metaphor for purity and the immaculate conception. It was used as an emblem of both Christ and Mary: the one miraculously born by virgin birth,

---

* The pearls were described as the two largest ever seen. After Cleopatra's death, following her defeat by Caesar, the other pearl was said to have been split into two and used to decorate the ears of the statue of Venus – the goddess whom Caesar claimed as his ancestor and the Roman equivalent to the Greek Aphrodite – at the Pantheon in Rome.

the other the chaste mother herself.[24] The two associations suited the Queen, portraying herself as the virtuous new saviour, and nurturing mother, of her country.[25]

But there was much more to her pearls than just this. The most famous paintings of the Virgin Queen show her bedecked in pearls, as snow-white as her leaden made-up face. This was a power move, and one with a very feminine dimension. The Armada Portrait of 1588 that commemorated the destruction of the Spanish invasion force is a panoply of pearl. The Queen wears five strings of massive pearls, her cloak is fringed with them on all sides and two dozen are arranged around her hair and headpiece. In the background sits the Tudor Crown, itself rich with pearls, and on the walls are paintings depicting the fate of the Spanish fleet.

In a portrait that does not wear its symbolism lightly, with the Queen's right hand resting on a globe, the pearls make a series of very clear statements. As it had in ancient Rome, the pearl signified status, especially in such rich abundance. Yet in the context of the Elizabethan era, an age of competing maritime empires, the sea-born pearl was also a symbol of power, and a very direct allusion to the triumph of English naval might over the Spanish. The pearl had been a central feature of Spain's sixteenth-century empire, flooding into Europe in extensive quantities from fisheries off the coast of modern Venezuela, which Columbus encountered during his third voyage of 1498 (and which, along with Spanish trading ships, were the subject of raiding by English privateers).[26] Now the Queen of England was lauding her victory through the medium of pearl – the very gem that had helped to define a century of Spain's maritime supremacy.

No effort had been spared in wringing symbolic value from the pearl and, for the Elizabethans, there would have been no question what the pearl represented in the Armada Portrait. The painting also depicts a large pearl held below the Queen's waist, hanging suggestively from a tied pink bow. This served not just to symbolize her virginity, but to connect it directly to her power, making the Queen's very body an allegory for a nation that had repelled invasion.[27]

So closely tied to Elizabeth I was the pearl that it became a way for courtiers to curry favour and establish allegiance. A portrait of Sir Walter Raleigh, also dating to the aftermath of the Armada in 1588, shows him bedecked in pearls – decorating his cloak, buttoning his doublet, and most prominently worn as jewellery, an earring from which hang two huge, drop-shaped pearls. Through this depiction, Raleigh exhibited his connection with the Queen, as well as his privateering exploits to seek out the treasures of the New World.[28] Pearls were also an important gift from the man Elizabeth came closest to marrying, her favourite courtier Robert Dudley. On his death, also in 1588, he bequeathed her gifts including 600 'fair white pearls'. The Queen kept his last letter by her bedside until her own demise, fifteen years later.[29]

Elizabeth I's pearl-draped monarchy symbolized how the timeless virtues of the pearl were increasingly being shaped into statements of female power – in a century where women ruled not just in England but France (Catherine de' Medici) and Spain (Isabelle I of Castile). In the closely entwined royal dynasties of the era, these women even ended up wearing and competing over some of the same pearls. Catherine made gifts of pearls to her daughter-in-law, Mary Queen of Scots, on her marriage to Francis II in 1533, from the rich sets of pearls that Catherine herself had received for her own marriage to the Dauphin, the future King Henry II. When Mary's jewellery collection was put up for sale after her imprisonment in 1567, Elizabeth succeeded in acquiring the pearls, despite attempts from Catherine to reclaim her former possessions and assurances from the French ambassador that she would be able to do so.[30]

This is a far from unusual story in the history of pearls, whose most famous examples have been through complex chains of ownership stretching across centuries. One of these, another pearl with a royal lineage, would undergo a particularly momentous journey, connecting the identity of the pearl as a vessel of power and propaganda to its modern status as an icon of fashion and beacon of glamour.

IN 2011 one of the most precious pearls the world has seen came up for sale at auction, almost 500 years after it had been fished from the Gulf of Panama. This was La Peregrina – a near-perfect pear-shaped pearl that, at 2.5 centimetres in length, fitted perfectly in the palm of my hand. It is difficult to describe the effect of handling such a gem: the purest hit of pearl it is possible to imagine, bright white, symmetrical in shape, and a surface so flawless, a lustre so sharp that I could practically see my smiling face reflected in its sheen.*

That sense of immediate appreciation is one I have felt more

---

* La Peregrina is not to be confused with another famously beautiful and similarly named pearl, La Pelegrina, also often said to have been discovered in Panama and exported to Spain in the sixteenth century, from where it entered the French Crown Jewels. Although this origin story is unsubstantiated, it did famously belong to the aristocrat Felix Yusupov, one of Rasputin's assassins. To add to the muddle, there is another pearl known as Mary Tudor's Pearl, long confused with our pearl here, and whose history as a wedding gift from Philip II of Spain to Mary I of England has erroneously featured in many of the back stories of La Peregrina.

than once, when handling the most precious pearls that have crossed my desk. With coloured gemstones and diamonds there are so many details to dissect – cut, clarity, colour – but, with the pearl, its full beauty hits you straight away. Examining Marie Antoinette's pearl and diamond pendant, another enormous drop which came up for sale seven years later, I wrote only one word in my catalogue: WOW. It was not only huge, hefty and immensely touchable (all I wanted to do was keep holding on to it), but its provenance, through the French Queen's descendants, was perfect. Boosted by its connection to one of history's most famous women, it sold, against an enticing auction estimate of $1–2 million, for a record-breaking $36 million.

The Marie Antoinette association is highly prized, with stories abounding of how some of her most precious jewels were smuggled out of France as she was imprisoned, awaiting her trial, and her ultimate execution. Yet it is not a guarantee of a sale. In 2007 I was part of a team offering a necklace mounted with blood-red rubies and some of her alleged pearls. When it failed to sell (against, admittedly, quite a high estimate), the narrative developed that buyers were reluctant to string around their neck the pearls of a woman who had so famously lost her head.

The effect of a premium pearl is more than just visual or aesthetic, it is physically instinctive, as if inviting you to touch it. As Elizabeth Taylor, the last in a line of the Peregrina's famous owners, described it: 'The pearl was so tactile. I couldn't stop rubbing it.'[31] Lord (Frederick) Hamilton, whose family had owned the pearl for much of the nineteenth and twentieth centuries, similarly related how 'on the rare occasions when [it] came out of the safe, I loved to stroke and smooth its sleek, satin-like sheen.'[32]

The Peregrina's story started in the sixteenth century, when it was fished out of the waters off the coast of Panama in 1577 by a young African slave. The discovery earned the boy his freedom, and launched the story of what would become the most famous pearl in the world. The pearl was sent to Spain (which had colonized Panama), put on public display in Seville, and christened 'La Peregrina'. Often mistranslated as the 'Wanderer', its nickname was more likely a reference to its rarity, or its 'exotic' or 'unusual' origin: a 'Pilgrim' in the truest sense of the word, coming from abroad to foreign lands. It was purchased for an immense sum by Philip II of Spain, beginning several centuries as part of the Spanish Crown Jewels.[33]

By now, tensions were escalating between England and Spain, in part due to the Protestant–Catholic divide, but also due to smuggling and privateering by the English under Elizabeth I. The result was the Anglo-Spanish War, culminating in a deadlock, despite the English defeat of the Armada in 1588. Given the political symbolism of the pearl for both maritime powers at the time, it was entirely apt that Queen Margaret of Spain (the wife of Philip III) chose to wear the Peregrina at the signing of the peace treaty between the two nations in 1604.

It wasn't until nearly 200 years later that the Peregrina left the Spanish ruling dynasty for the French, when Napoleon forced the abdication of Ferdinand VII, and handed the Spanish Crown (and its treasury) to his elder brother, Joseph Bonaparte, in 1808. Joseph absconded with the pearl at the restoration of Ferdinand VII in 1813, and left it in his will to his nephew, the future Napoleon III who, in financial difficulties and in exile in England, sold it to James Hamilton, the 1st Duke of Abercorn. It was in the ownership of the duke's wife, Louisa, that the famous pearl won its reputation for wandering in a different sense. As their son Frederick later related, the Peregrina had never been drilled to allow it to be strung, and its significant weight meant it had a propensity to fall out of its setting. 'To my mother it was an unceasing source of anxiety,' he recalled, describing how she had once lost it at a ball at Buckingham Palace only to discover it caught in the velvet train of another woman's dress, and how on another occasion it went missing at Windsor Castle, this time to be found down the back of a sofa.[34]

Despite these mishaps, the Peregrina would remain in the Hamilton family until 1969, when it was sold, through the trade, at auction. The buyer was Richard Burton: he had paid $37,000 and, in a faint echo of old enmities, had seen off a bidder who said he was related to the Spanish royal family and wished to return the prized pearl to its rightful home.[35] The Peregrina thus became part of the Burton–Taylor love story, which is how it came into my hands when Elizabeth Taylor's collection was sold by Christie's in 2011. The gem would be bought for $11.8 million, at the time the most expensive pearl ever sold at auction.

No sooner had Elizabeth Taylor been presented with the jewel in 1969 than she too had lost it – proving that time had not dimmed this pearl's slippery reputation. Discovering it had disappeared off its chain in their penthouse at Caesars Palace in Las Vegas, she went down on her hands and knees, desperately hoping that her husband – who was 'in one of his Welsh moods' – would not realize what had happened. Suddenly she saw one of their two Pekinese dogs chewing on what appeared to be a bone, but realized that they never gave the dogs bones to chew on. 'I just casually opened the puppy's mouth, and inside its mouth was the most perfect pearl in the world. And it was – thank you God – not scratched.'*[36]

---

* It wouldn't have been the first famous pearl ingested by a pet: the American heiress Barbara Hutton supposedly fed the pearls from her magnificent necklace – also said to have belonged to Marie Antoinette – to her goose, so that 'they would come out with a brighter lustre'.

The Peregrina had survived its haphazard journey down the centuries – from peace treaties to puppy treats – much as the pearl's own associations of beauty, status, purity and aspirational style remained constant. Yet the place of the pearl in society was to become more dynamic, driven by social change and market forces. In the twentieth century, with luminaries such as Elizabeth Taylor to the fore, the pearl was cast in a fresh light, and brought to a wider audience. As the jewellery historian Vivienne Becker has suggested, the pearl was carried along by a sea change of social mores in the 1920s, as the twentieth century truly dawned. 'Women now worked, drove, drank and smoked and danced in public, played sports and, most important, bought their own clothes and jewels. Most often, they wanted to wear styles they had seen in the movies, aspiring to previously unattainable luxury and glamour.'[37]

One designer above all helped to shape this new era of style: Coco Chanel. She was the primary inspiration behind the slim silhouettes, short hair and angular accessories that marked independence from the restrictive fashions of the past, for a new generation of newly independent women. With Chanel came the age of the little black dress: practical, universal and devastatingly chic. And with this 'flapper' uniform came another Chanel signature: pearls in abundance, 'ropes and ropes' of them that could wind around the neck several times.

The pearl was a natural fit for the pared back, Art Deco aesthetic, and one which crossed the classes. Chanel's own characteristic pearls were valuable natural gems that were part of a jewellery collection including gifts from the Duke of Westminster and the Grand Duke Dmitri Pavlovitch (a cousin of the Tsar),[38] while flapper pearls could be a simple fashion choice: indistinguishable cultured, or even imitation, gems. This was entirely in keeping with the desire to cater for the modern woman: to revel in the freedom of no longer being judged by outdated standards. As Chanel herself put it: 'I love fakes because I find such jewelry provocative, and I find it disgraceful to walk around with millions around your neck just because you're rich. The point of jewelry isn't to make a woman look rich but to adorn her.'[39] Even the linguistic element of the product reflected this progression: to the Romans, the pearl was 'unio', named after its uniqueness, while still today, in French, the word 'perle' can be used interchangeably for an oyster pearl or a plastic bead.

This democratizing trend gathered even more momentum following the Second World War, as the production of cultured pearls from Japan resumed and rapidly expanded – between 1951 and 1966 there was a forty-seven-fold increase in the number of Japanese pearl farms.[40] This tide overwhelmed the natural-pearl industry's attempts to restrict and rubbish its new competition; even jewellers who had initially held out, including Tiffany's, were now stocking cultured pearls.*[41] As a result, the style icons of the era – from Elizabeth Taylor to Marilyn Monroe and Grace Kelly – were increasingly advertising a product that any woman could own. When Audrey Hepburn starred in Breakfast at Tiffany's (1961), accessorizing her black Givenchy dress with five strings of fat pearls, it had never been easier for the woman sitting in the cinema to replicate the look.

The twentieth-century glow-up of the pearl reflected the

---

* The industry had brought a legal case in the 1920s, hoping to have cultured pearls declared fraudulent. Instead, a French court judged that the new varieties 'do not differ . . . at all'. The natural-pearl market also took a dive in the wake of the Great Depression, which saw prices tumble and the US dealer network hollowed out – from several hundred in the 1920s to a handful by the 1950s.

particular qualities of a gem that has been in human hands longer than any other. It is a timeless classic that has also been the subject of many subtle reinventions – different ways to wear it, new ways to produce it, associations that have evolved as well as endured. The pearl's aesthetic simplicity means it provides a blank canvas for each generation to redefine, while its timeless symbolism creates a deep meaning that conveys value on even the cheapest imitation. It will perennially be the gem that everyone adores, that suits all looks and situations, and which essentially advertises itself. 'Pearls are always appropriate,' said another 1960s trendsetter, First Lady Jackie Kennedy. Famously she had purchased her own imitation pearls for a mere $35 from Bergdorf-Goodman, the New York department store.

The style icon and her cheap pearls perfectly sums up the range of this gem: one that has reached into the highest echelons of wealth and power at the same time as becoming accessible and affordable to all. It is the embodiment of the democratization of jewellery, women's liberation in gem form. So much has changed in 7,000 years of our history with the pearl. Once hunted as nature's rare gift, it is now cultivated and manufactured as humanity's product. Historically an elite symbol, it has become the gem that every woman can buy, whether from an historic estate for $35 million or a department store for $35. Yet for all this change, really nothing has changed about why we wear pearls. They are still the miracle of nature that emerges fully formed, with a shape, life and lustre that seems heaven-sent. A gem you want to wear all the time, to touch and feel – to hold and never let go.

6
Spinel

The Gem of the Mughals

*'Wine is molten spinels and the decanter
is the mine.'*

THE RUBA'IYAT OF OMAR KHAYYAM,
ELEVENTH CENTURY

W HEN I HAD last visited the Tower of London, it was surrounded by a sea of deep-red handmade poppies, one for each of the British or Colonial soldiers who lost his life in the First World War. My great-grandfather had been one of them, and his name was read out in the roll call. I was now back again, several years on, and this time the grey fortress was twinkling in the low yellow light. It was just before Christmas, and I was in search of something very special: one of the most magnificent yet misunderstood gems in the world, the Black Prince's Ruby.

Set full-frontal on the Imperial State Crown of the British monarchy, above a piece of the largest diamond ever mined and below a sapphire thought to have been worn by the last Anglo-Saxon king, this gigantic gemstone hides in plain sight. As the second-largest stone of the thousands that decorate the world's most famous headpiece, the Black Prince's Ruby is hardly lacking in stature. Its deep, saturated red beams out of the most recognizable royal images, including Cecil Beaton's coronation portrait of Queen Elizabeth II, where she sits crowned in Westminster

Abbey, the sovereign's sceptre in her right hand and orb resting in her left. Yet despite this prominence it is not especially well recognized, and even less well understood. Few could tell you what it is called and even fewer what it actually is. Absolutely no one can tell you with total confidence how it got there.

While the diamond below it glitters with its cushion-cut brilliance, this stone has no facets. Polished down from the original rough, its uneven, indented surface and lopsided form contrasts with the careful geometry of the crown's other jewels. Its huge, 170-carat size and rich, purplish-red colour call to mind a squashed plum, or a massive jellied candy. It is a majestic gem, but seemingly out of place, as exotic and surprising as a peacock in an English country garden. In context it can appear almost random, as if added on by mistake or as an afterthought.

This is just one of many misapprehensions it is possible to have about the Black Prince's Ruby, a gemstone surrounded by almost as many myths and layers of confusion as it has carats. Despite its rough-and-ready appearance, it has been a venerable mainstay of the Crown Jewels, present for much longer than the exquisitely fashioned Cullinan Diamond below it. It dates back at least to the Crown of State worn by King Charles II at his coronation in 1660, and possibly some time before that. Its name has stuck despite the fact that it almost certainly had nothing to do with the Black Prince, the eldest son of King Edward III, and is definitely not a ruby. And to confuse things even further, the stone that is named 'ruby' while not actually being one does, bizarrely, *contain* a real ruby. Noticeably pinker, this small stone within a stone is set in a thick gold collet, filling a drill hole originally made for the mother gem to be strung and worn on a necklace, headpiece or armband.

This hodge-podge of high quality, bad branding and problematic provenance makes the Black Prince's Ruby a fitting symbol of the gem family it actually belongs to – spinel. A much misunderstood, often misrepresented gem type, spinel has nevertheless played a starring role in the royal treasuries of India, Russia, Iran, Turkey, Britain and France since the Middle Ages. For centuries the spinel enjoyed a distinct status as a jewel of the highest rank, at points being esteemed above even ruby and diamond.

Yet despite the spinel's rich heritage as a cultural icon, it has not in its later life enjoyed a reputation to match. Today, the most famous spinels are not even known by their own name, but carry that of another gemstone: the 398-carat whopper atop the Great Imperial Crown of Russia is called Catherine the Great's Ruby, and the huge cabochon that adorned the Peacock Throne of the Mughal emperors in the seventeenth and eighteenth centuries is labelled the Timur Ruby.

These misnomers are indicative of how uncertainty came to settle over a gemstone that had been prized by the world's most powerful, but latterly came to be considered as somehow an impostor, pretending to be something it was not. All gemstones have to some degree seen their fortunes fluctuate and their reputations rise and fall through history – driven by a combination of fashion, supply and demand, and the emergence of competitor stones. But none has endured so dramatic or unfortunate a journey as spinel, which went from being the toast of Mughal and Persian emperors to a gemstone that lost not only its lustre but its very identity, subsumed by the increasingly popular and prevalent ruby – whether being mistaken for it or regarded as its poorer cousin.

For the medieval Arab lapidaries who first wrote about the spinel, there was no mix-up between the two red stones: la'al (spinel) and yaqut (the catch-all term for ruby and sapphire).[1] They had unconsciously anticipated later discoveries showing that the two minerals are distinct both chemically – spinel as magnesium aluminium oxide ($MgAl_2O_4$) to ruby's simpler aluminium oxide (Al_2O_3) – and in their crystallography, with spinel's cubic structure having more in common with the diamond than ruby. Its octahedral shape, culminating in two pyramidal points, is the most obvious explanation for why it was dubbed *spinelle* (Latin *spina*, thorn).

Spinel was the most beloved gem of the Persians and Mughals, rightly prized for its attractive colour, high clarity, and abundant carat weight. Huge, polished beads did not need to be cut, and offered an extensive surface area to be engraved with the names of emperors and their ancestors, a family tree cut in crystal as a perpetual record of dynastic power. In its sixteenth- and seventeenth-century heyday, spinel was a gem at the heart of the Renaissance in the East, paraded and prized by royalty, at courts famed for their lavish culture, architecture and embrace of the arts.

By the late nineteenth century, as rubies began to emerge commercially from Burma and steal the show, spinel had become relegated to second-class status, regarded at best as an inferior red stone and at worst as something akin to a fake. European collectors either failed to recognize or declined to appreciate the distinguishing qualities of the spinel that had so engaged its Eastern admirers, who esteemed it as both distinct from and superior to its fellow red gem. For the Mughals and Persians, spinel had been the gemstone of royalty for good reason: a rich red tinged with pink and purple that was a favoured metaphor of Iranian court poets, and a stone big and tough enough to engrave with dynastic claims. The truly showstopping examples, like the Black Prince's Ruby, remained royal gemstones. But more broadly, what had been highly prized *la'al* in the East increasingly became underrated 'spinel ruby' or 'balas ruby' in the West. A stone once ranked above diamonds came to be regarded as little better than a second-rate ruby. Spinel truly is a gem that has lived a double life.

In the course of that haphazard journey, spinel has tended to attract stories wild and wonderful even by gemstone standards. Confusion and misattribution have stalked not only the spinel family as a whole, but many of its most famous examples. None more so than the Black Prince's Ruby, a gem whose history is as wonky as its appearance, full of tenuous associations but light on verifiable evidence.[2] The narrative attached to it begins in the fourteenth century, with gifts given by Peter I of Castile to the Black Prince, in recognition of English military support in his long-running succession struggle – a proxy conflict that helped to reignite the Hundred Years War between England and France. The elusive jewel crops up again in the fifteenth century, during the reign of Henry V, with an almost certainly mythical description of a 'ruby' mounted on the crown he wore over his helmet at the Battle of Agincourt in 1415, the stone said to have stayed resolutely in place, and even saved the king's life, when part of the crown was hacked away by an axe. More credible is the record that part of a crown featuring a large red stone was pawned the same year (prior to Agincourt) to help cover the costs of the campaign. References become more concrete in the sixteenth century. Henry VIII's crown featured a prominent red gem described as broken, matching the naturally flattened form of the back of the Black Prince's Ruby.[3] The 'Ditchley Portrait' of Queen Elizabeth I, dating to around 1592, shows her wearing a headpiece topped with a red gem that appears to contain a drill hole similar to the current stone. We can be reasonably confident that the 'Wonderful large Ruby, set in the middle of one of the four Crosses' on the crown of Charles II is the gem we know today. That is reinforced by the first piece of convincing visual evidence: a 1718 painting of George I's Crown of State, whose centrepiece gemstone is recognizably the wobbly pear now set in the Imperial State Crown.

It is clear enough that the Royal Family have been using crowns adorned with large red gemstones since at least the fifteenth century, but there is no certainty at all that these were one and the same jewel. The Black Prince's Ruby may have been the '*grosse escarboucle*' (large carbuncle) that decorated an ornate table given by Philip I to Prince Edward in the 1360s. It might have been the spinel offered for sale to Henry V in 1414, when he was designing a crown for his marriage to Catherine of Valois. It could have been the 'large and very precious' red stone presented to Catherine of Aragon a century later, or the 'large Orientall ruby' bought by Charles II prior to his coronation in April 1661. We will probably never know for sure.

Given all this uncertainty, it is fitting that the Black Prince's Ruby earned its name in a case of double mistaken identity, in part perpetrated by the eighteenth-century Whig MP and historian Horace Walpole. Shown a painting of a man labelled as the Black Prince – but whose identity was widely disputed at the time, as it has been since – Walpole took particular interest in the headgear: 'He has a hat with a white feather, and a large ruby, exactly in the shape of the rough ruby still in the crown,' Walpole wrote. Although neither his identification of the man in the portrait, nor the gemstone on his hat, appears to have been sound, the idea caught on and the 'Black Prince's Ruby' was born.[4]

This is just one example of how famous spinels have come to be saddled with incorrect associations. By contrast, the gemstone's role through history has been reasonably clear – the combination of size and colour making it fit for Crown Jewels and the most powerful of patrons. Comb through the treasuries of history's richest dynasties, and the red stones of note are invariably spinels. Only comparatively recently did the spinel start to be confused with and pejoratively compared to the ruby, earning its disobliging moniker as the 'Great Pretender'. But the real story here is of the Great Misnomer: a gem that has variously suffered for being misnamed, misidentified and misunderstood. Strip away the mistaken identity, and the radiant glory of this rich red gemstone starts to emerge once more.

SPINEL'S LATTER-DAY reputation as the ruby's poor relation would have come as a surprise to the collectors of earlier eras. Early documentary evidence makes clear that this was regarded as a very precious gemstone indeed. 'There is but one special mountain that produces them,' wrote the traveller and merchant Marco Polo (1254–1324). 'The stones are dug on the king's account, and no one else dares dig in that mountain on pain of forfeiture of life as well as goods; nor may any one carry the stones out of the kingdom.'[5] The closely regulated gem to which he referred was *la'al-e Badakshī* – literally something red (Arabic *la'al*) from Badakhshan. *La'al* (*lāl* in Persian) was the name given to the very particular shades of red that are characteristic of spinel – the best stones being a deep red, but others appearing in softer pinks or purples.* Badakhshan was the historic region encompassing parts of modern Afghanistan and Tajikistan, and home to the mines of Kuh-i-Lal, one of the most significant producers of large red and pink spinel in the world.†

---

* The Iranian art historian Assadullah Souren Melikian-Chirvani shows how *lāl* was the etymological root in Persian for other reds, including *lāle* for the wild anemone flower.
† Much has been made of the historic mines of Kuh-i-Lal in Tajikistan in recent years; it has become the best known, but it is by no means the only spinel deposit in the area. Several mines are located in both Tajikistan and Afghanistan along the border with a few contenders for the original 'special mountain' of Marco Polo.

Sometimes written as 'Balakhshan', this was the source of the later European tendency to label spinel as 'balas ruby', the beginning of the slippery slope that would lead to much later misattribution.

The discovery of spinels in Badakhshan is thought to have followed a major earthquake at some point in the seventh century. 'It is said that the mine was located when there was an earthquake in the area and the mountain was cloven,' the Persian scholar al-Biruni recorded. 'Big rocks fell down and everything was destroyed. *Laal* were disgorged in the process.' Initially, he suggests, local women tried to grind down these purplish-red stones as a form of dye.* When they were found to be useless for this purpose, they 'showed the rubies to men and the matter was publicised'.[6] Marco Polo was just one of many who later picked up the threads of this story: the earthquake that had brought these sparkling stones to life, the local women who had turned their noses up at them, and the royal monopoly that had quickly developed around mining them. Badakhshan would be the source of all the famous, historic and massive spinels we know today, 'sparkling pink gems of the size of eggs', as one nineteenth-century source described them.[7] As such accounts suggest, size was critical to the appeal of *laal*. No other source of spinels has ever produced crystals to rival these mauve monsters. The egg-sized gems that had been cast aside by Badakhshani washerwomen would go on to decorate the crowns of the world's most powerful dynasties, making Kuh-i-Lal and its neighbours one of those astonishing gem deposits that has defined the success and history of a gemstone.

---

* This is not as strange an idea as it may seem. The vibrant blue gem lapis lazuli, famously mined not too far away from Kuh-i-Lal in Sar-e-Sang, is a proven source of precious blue pigment.

Of all the spinel's fans in high places, none were more enthusiastic than the Mughal emperors who ruled large swathes of the Asian subcontinent at the height of their power in the sixteenth and seventeenth centuries. The Mughal Empire was a cultural and artistic as well as political powerhouse: its patronage of artists, architects and craftsmen was legendary, and the modern image of India largely remains its creation: from the Taj Mahal to Delhi's Red Fort and the Buland Darwaza (Victory Gate) at the dynastic capital of Fatehpur Sikri. The Mughal penchant for lavish art, vast architecture and showpiece jewellery was not simply the expression of wealth and power. It also represented the projection of identity. As Mongol–Turkish outsiders to what they called Hindustan, the Mughals strove to establish legitimacy by emphasizing their connection to a previous ruling dynasty: the Timurids, whose empire spanning Turkey and Central Asia had extended into modern Pakistan and northern India. Timur, also known as Tamerlane (1336–1405), a descendant of Genghis Khan, had captured and sacked Delhi in 1398. His great-grandson, Babur, who had inherited a tiny piece of the fragmented Timurid Empire, would repeat this feat in 1526 and become the first Mughal Emperor, reigning until his death in 1530. Babur, a warlord who wrote poetry almost as freely as he drank wine, and who seems to have enjoyed designing palaces as much as planning invasions, created the template for a ruling dynasty of aesthetes.[8]

Babur's successors amplified his taste for the extravagant as much as they expanded his physical empire. As the decades passed, Mughal emperors became renowned as the world's most lavish jewellery collectors, with extravagantly sized spinels at the forefront. The fourth emperor, Jahangir (r.1605–1627), was described by the English priest Reverend Edward Terry as the 'greatest and richest master of precious stones that inhabits the whole earth', and by the Flemish diamond merchant Jacques de Coutre as 'looking like an idol on account of the quantities of jewels he wore, with many precious stones around his neck as well as spinels, emeralds and pearls on his arms, and diamonds hanging from his turban'.[9] While many gemstones were embraced at the Mughal court, spinel held a special status. A catalogue describing the court of Jahangir's father, the long-reigning Akbar the Great (r.1556–1605), described how one of his treasury's twelve sections was reserved for gemstones, and that these were split into three groups: the first for spinel, the second for ruby, sapphire, diamond and emerald, and the third for pearl.[10] According to the Mughal expert Susan Stronge, 'all precious stones were eagerly sought, but spinels had a superior position and were kept separately'.[11]

In addition to its size and striking colour, the basis of spinel's pre-eminence can be explained by the importance of Persian tradition in Mughal society. Persian polymaths did not just sprinkle spinel into their poetry as a piece of imagery. They also approached it scientifically, the scholar al-Biruni describing *la'al* as 'a red gem, translucid, limpid, which resembles a superlative ruby in colour. It often surpasses it for its beauty and glamour but it differs from it in hardness'. These characteristics made spinels a regular adornment to the turbans and crowns of Persian rulers, where they often sat alongside rubies (the distinction underlining how early eastern connoisseurs of *la'al* had no trouble telling the two red gems apart). As the art historian Assadullah Souren Melikian-Chirvani has explained: 'spinels . . . appear to have shone with a unique brilliance in Iranian culture and later in Hindustan where Persian prevailed as the language of literature and social intercourse'. He has shown how the spinel features as a consistent metaphor in Persian poetry dating to the tenth century, with wine often being described as 'molten spinels', and '[the] ascent of the sun in the sky . . . likened to a spinel coming out of its mine'.[12]

The associations with sun, fire, blood and wine are familiar ones for red stones, and in different contexts have also been attached to both ruby and garnet. Yet in Mughal India, where Persian was the court language and 'courtly culture . . . drew on the rich and highly developed literary and scientific traditions of Iran', spinel was esteemed as first among equals.[13] Practical considerations also helped to elevate it. Its sheer size made it appropriate for the tradition of gemstone carving that was popular in Iranian and Mughal royal circles. The beads would be drilled with diamond-tipped points so they could be strung into necklaces and headdresses or strapped onto armbands, and they had the names of their past and present owners carved into them, along with notable dates and sometimes longer inscriptions.

Several spectacular examples have survived, in some cases assembled by later jewellers. One such necklace, featuring eleven polished spinels, sold at auction in 2011 for close to $5 million. Three of the stones are engraved, two with Jahangir's name, and a third also with his two successors: Shah Jahan (r.1628–1658) and Aurangzeb (r.1658–1707). All three names also feature on a 133-carat spinel bequeathed to the Victoria and Albert Museum in 1922 by Lady Carew.* With succession a continually contested issue for the Mughals, stones like these would have been valuable symbols used by emperors to assert their authority and

---

* That the Black Prince's Ruby is not engraved suggests it came to Europe relatively early, and may never have been owned by an Eastern royal dynasty.

legitimacy.* Their engraving was a way of validating the heredi-
tary record on something considered immutable and eternal:
hard, huge and historic precious gems.

Shah Jahan's life would end in ignominy, usurped and
imprisoned by his son Aurangzeb, but at the height of his power he
had been the archetypal Mughal ruler, responsible for some of the
empire's most lasting icons. He consciously presented himself as a
Timurid – choosing to wear a full beard as the dynasty's most famous ancestor had done, unlike his father or grandfather, and
adopting a title that had also been attached to Timur, styling himself as the second 'Lord of the Conjunction'.[†14] He was also a builder
on a scale grand even by Mughal standards: the architect of both
the Taj Mahal, tomb for his beloved wife Mumtaz Mahal, and the
Peacock Throne, a beyond blinged-up stately seat that cost twice as
much as the Taj had done and took seven years to complete.

Sitting on top of the Peacock Throne were two gemstones
that would become enduring symbols of power and conquest in
the centuries that followed, as Iranian, Sikh and finally British
forces took possession of the supreme Mughal treasures. One
was the Koh-i-Noor diamond. The second, an engraved spinel,
was the stone that became known as the Timur Ruby. Like the

---

* Aurangzeb had a particular need to stress legitimacy. He had fought a war
of succession against his older brother, and after securing victory had him
put to death and their father (Shah Jahan) imprisoned for the final eight
years of his life. By carving his name onto the spinel already featuring those
of his predecessors, he suggested a far cleaner succession than had actually
been the case.

† The title referred to the astrological event in which Venus and Jupiter come so
close to each other that they appear to collide, said to have occurred when
Timur was born. Shah Jahan was not the only ruler to subsequently appropriate
the label, with important consequences in the history of one famous spinel.

Black Prince's Ruby, this spinel is now in the British Royal Collection. And like that stone it carries a double misnomer, for this cabochon is neither a ruby nor was it ever owned by Timur. It is another example of spinel's deliciously twisted history, and the layers of confusion that have built up around these enormous stones. As is so often the case with spinel, a large gemstone has given rise to a series of even taller stories.

Weighing 352.5 carats and shaped like a slightly squashed human heart, the Timur Ruby is engraved with the names of five Mughal emperors, along with that of Nader Shah, the Persian prince who had taken the gem – along with the Peacock Throne in its entirety – when he sacked Delhi in 1739. The British seized it from its Sikh owners after their annexation of Punjab in 1849, along with treasures including the Koh-i-Noor. In her journal, Queen Victoria recorded her admiration for the gem: 'The rubies [sic] are even more wonderful [than the pearls and emeralds]. They are cabochons, unset but pierced. The one is the largest in the world, therefore even more remarkable than the Koh-i-noor!'*[15] This special 'ruby' was then mounted into a diamond necklace with three smaller spinels by Garrard in 1853.

Confusion crept in over half a century later, when Sir James Dunlop Smith, then private secretary to the Viceroy of India, was deputed to source the provenance of several Crown Jewels of Indian origin, in response to a request from Queen Mary, the wife of George V. Susan Stronge, who has studied the Timur Ruby in detail and done more than anyone to sort truth from abundant myth, has shown that Dunlop Smith's work set off a

---

* In fact, other famous spinels outweigh the Timur, including Catherine the Great's Ruby (398.72 carats) in the Great Imperial Crown of Russia, and the 500-carat Samarian Spinel that is part of the Iranian Crown Jewels.

train of misunderstandings, mistranslations and misapprehensions that forged one of the great gemstone myths.

Dunlop Smith was given special access to the Timur Ruby, which was removed from its mount so he and a translator could examine the inscriptions. Yet despite this close examination, he appears to have confused it with another spinel he had read about that had belonged to Timur and which was also engraved with the 'Lord of the Conjunction' title. Rushing to make the association, he appears to have glossed over inconsistencies in the inscriptions, as well as the fact that this title was adopted by multiple emperors, including not only Timur and Shah Jahan, but also Nader Shah. We now know that it was to the latter that the inscription almost certainly referred, but Dunlop Smith did not. The upshot was yet another misidentification of a famous spinel. Like Horace Walpole before him, Dunlop Smith had looked at a beguiling red stone and found what he wanted to see. Thanks to him, Victoria's 'ruby' would be for evermore known (in every sense incorrectly) as the Timur Ruby.[16]

By this point, it was not only fake history that was starting to bedevil the spinel. The item was losing not just its authentic provenance, but its entire identity as a gemstone. By the time the Timur Ruby was being misattributed in the 1910s, spinel had come to be defined in unflattering contrast to the ruby and as a subset of it. A gemstone that had been the most prized of all was increasingly discounted as something of little worth, a poor imitator of the ruby to which it had once been deemed preferable. As is often the case with bad reputations, the spinel's was too easily earned, and would prove incredibly difficult to shake off.

IN 1892 the British jeweller Edwin Streeter was publicizing a new edition of his guide to precious stones, and turned his attention to the subject of fakes. The *Daily News* reported him sharing the story of a customer who had brought family heirlooms to him for valuation, only to be told that the blue stones purporting to be sapphires were in fact worthless imitations. 'But paste is not the only substitute for real gems,' the report continued. 'The Spinel and the Balas, the one a lively poppy red, the other a violet rose, frequently usurp the dignity of a true Ruby . . . the pure Ruby of ten carats is almost beyond valuation, while the other stones, called by the same name, are only of trifling value.'[17] How had a gemstone once admired for its distinctive properties, esteemed and valued above the ruby, become a dirty word in the jewellery trade?

One significant factor was the emergence of the science of gemmology, which subjected gemstones to increasingly detailed scrutiny that allowed for finer distinctions to be made. As the study of gems became more professional, the reputation of the spinel seemed to sink with it. In 1783 the pioneering French crystallographer Jean-Baptiste Louis Romé de l'Isle established the difference in crystal structure between the trigonal ruby and the cubic spinel, using new equipment to measure the angles along different faces precisely. Then in 1812 the German mineralogist Friedrich Mohs published his scale of mineral hardness, confirming that ruby and spinel also differed on this measure: ruby was 9 on the scale, while spinel was at 8. Whereas spinel's earlier Eastern admirers had esteemed the stone for its core attributes of striking size and carmine colour, the early Western gemmologists increasingly highlighted the same gem for all that it was not: those same things that made it different from, but in their eyes lesser than, the ruby.

At the same time as spinel was being distinguished empirically from ruby, it was also being lumped together with it in its nomenclature. Initially, early references to 'balas' in the thirteenth century were quite distinct.[18] When variations of the term 'spinel' were introduced in the sixteenth century, they referred to stones now coming from Burma, while the 'balas' remained what the Persians had dubbed *la'al-e Badakshī* (the red stone of Badakhshan). But by the nineteenth century, jewellers began to refer not only to 'balas rubies' but also 'spinel rubies' as varieties of spinel which, Streeter emphasized, should be 'readily distinguished from the true or Oriental Ruby, with which it has been sometimes confounded'.[19] The message was clear: ruby was the prince of red stones and spinel the pauper.

In the hands of European gem specialists,* spinel was suffering a dual misfortune: either it was being classified as an inferior ruby or it was actually being mistaken for one. With so many different names and terms floating about, it was only too easy for the material to be misidentified or wrongly described. What had been referred to as 'balas ruby' or 'spinel ruby' soon simply became ruby. Queen Victoria's reference to the spinels in her collection as 'rubies' suggests how widely this broad brush may have spread. With friends like these, spinel needed no enemies. Rather than being allowed to flourish in its own right – as *la'al* had done in the

---

* European taste at the time might also have favoured faceted stones over the ubiquitous smooth cabochon form of the Mughal spinels. One exception was the Hope Spinel, owned by the famous gem collector Henry Philip Hope (of the eponymous blue diamond fame), a spectacular 50-carat square step-cut spinel from Badakhshan which showcased the brilliance that a beautifully cut spinel could exhibit: when it emerged onto the modern market in 2015 it duly set a world record, selling for just under £1 million.

hands of its Mughal and Persian admirers – spinel was becoming entirely subsumed within the ruby's increasingly dominant brand.

Spinel's reputation was also a victim of circumstance in the late nineteenth century. In the 1880s and 1890s the craze for Burmese rubies was entering full swing: the British annexation of northern Burma in 1885–6 opened up the exploitation of the Mogok mines, backed by investors including the London bank of N.M. Rothschild. Excitement was intense: when their exploration company, Burma Ruby Mines Ltd, was floated on the London Stock Exchange in February 1889, such a large scrum descended in search of shares that Lord Rothschild had to climb up a ladder 'in a burglarious fashion' to escape the crowd and get into his office.[19] Although both spinel and ruby are found in the Mogok region (a coincidence that undoubtedly led to even more confusion), these nineteenth-century prospectors were only after one form of red gold, and that was ruby.

The spinels mined in Burma were not a patch on their cousins from Badakhshan: bright red at their best, but far from the same mighty size. Ruby was the prize in Burma, and Mogok became famous as the 'Valley of Rubies'. The weight of the jewellery industry swung firmly behind marketing ruby as the red stone of choice, with the strong support of one person in particular. Edwin Streeter was an opportunistic gem dealer and jeweller, an influential figure whose professional opinion was strongly sought after, but he was also one of the key financial backers of the Burma Ruby Mines Ltd. When he deprecated the spinel in comparison to the 'Oriental' (i.e. Burmese) ruby, he was speaking not just as a respected industry figure but as a major investor in Burma Ruby Mines Ltd, who may have been expressing his commercial interests as much as his professional opinion. It was paramount that ruby prices rocketed, and if that was at the cost of the once-hallowed spinel, so be it.

However they were motivated, these efforts contributed to a clear divergence in both the reputation and valuation of the two red stones. And it was not just marketing that made the difference. The mineralogical distinction between ruby and spinel only served to make the increasingly sought-after ruby rarer, and therefore more valuable. By contrast, spinel was falling from grace, a reality reflected by tumbling prices. In the 1892 edition of his study, Streeter cited an inventory of the French Crown Jewels from 1791, which valued a 56.75-carat 'spinel ruby' at £2,000 and a 20.38-carat 'balas ruby' at £400, before cautioning (one assumes again not without bias): 'it should be stated that at the present day the stones would not fetch one-tenth of such prices. Today,' Streeter continued, 'the Spinels are not much cared for.'[21]

That statement would become even more profound in light of the final loss of face in the spinel story. Just a few years earlier, French chemist Auguste Verneuil had discovered the flame fusion process, a method by which corundum – and soon, spinel – could be chemically and structurally recreated in a furnace. By the 1930s, synthetic spinel had become commercially available, easy and cheap to manufacture and in a range of different colours which were produced to imitate other gems.[22]* It would also become the mainstay of American class or graduation rings, so widely worn in the United States that the word 'spinel' would become a byword for 'fake' in some circles.[23] It looked like the final nail in the coffin for the once-illustrious spinel.

---

* One of the gems I would always test on principle whenever it came in for valuation would be light blue aquamarine – frequently set in large cocktail rings of the 1930s and '40s – in case it was a synthetic spinel.

The damage done by Streeter and synthetics would be long-lasting, but not permanent. If the history of gemstones teaches us anything, however, it is that no stone's story is ever entirely told, and changes in fortune may lie just around the corner. The field gemmologist and spinel enthusiast Vincent Pardieu has described how a new find was physically dropped into his hand in 2001, when he was working in Myanmar: 'stunning little gems with a bright neon pinkish red color, convincing me that spinels could equal rubies in beauty'.[24] These gems were unlike any spinels ever seen before: vivid pink octahedral crystals lacking any of the 'dark tone' seen in other examples. On that basis, and with a recent *Star Wars* film in the back of his mind, Pardieu christened them 'Jedi spinels', a name that rather appealed to the trade. Dosed up with high quantities of chromium, these hot pink pebbles came from Burma's Namya and Man Sin mines, which I later visited on my own trip to Mogok. I adored these little miracles of nature. Unlike the more rounded form in which Badakhshani spinels were historically found, the Jedis were perfect, sharp octahedral crystals, which already shone brightly when they were first pulled from the ground. It was not hard to see why the locals had dubbed such octahedra *nat thwe* – 'polished by the spirits'.

Propelled by smarter branding than the spinel had typically enjoyed, these Jedi spinels quickly made a splash on the Asian auction circuit. Then in 2007 another remarkable discovery followed. The unearthing of a 52-kilogram lump of rough spinel in Mahenge, Tanzania, signalled the discovery of a significant new deposit of bright pink-red gems. The region would subsequently yield not only more red and pink stones but examples of the extremely rare cobalt-blue spinel variety which had begun to be mined in Vietnam in the 1980s.[25] In the early twenty-first century, a stone that had become moribund and unloved started to enjoy a new lease of life. In a delicious irony, its fortunes have also been boosted by the stone that was once its nemesis. The exponential rise of ruby prices has broadened the market for coloured gemstones; spinel, with its wider variety of colours, and remarkable clarity that requires no heat treatment, has been a clear beneficiary. Ruby's poor substitute has started to become, in the right context, its viable alternative.

This twist in the tale continues a history of fluctuating fortunes, extreme even by gemstone standards. Other gems have gone from popularity to obscurity, from a plaything of the rich and famous to the toast of the cheap and cheerful. But perhaps none has travelled from such a high as spinel enjoyed at the court of the Mughal emperors to the low it touched at the turn of the twentieth century. Certainly no other gemstone has been saddled with such an unfair reputation as spinel, in the rush by the early gemmologists to distinguish it from ruby, with seemingly no thought to why this alleged impostor had once been regarded as such a precious prize.

The spinel story may be one defined by confusion and misapprehension, but it is also revealing about the forces that variously propel gemstones to prominence and bring them crashing back to the earth from which they came. Like the diamond, spinel shows how gems are products of marketing as much as mineralogy. In the century when diamond was the beneficiary of one of history's greatest advertising campaigns, spinel was suffering because its brand had become hollowed out, shorn of its former association with royal magnificence and fatally typecast as a symbol of fraud. And like every other gem, but perhaps to an even greater extent, spinel reveals how these stones are objects shaped by human aspiration. The most famous spinels have been stones onto which emperors have carved their names, with which rulers have sought to assert their legitimacy, and around which historians have woven often fanciful narratives. They are not just historical objects, but ones that have been used to shape and sometimes distort history.

While being used to tell the stories of their owners, the narratives of these stones have frequently become lost and confused in the process. Yet through that confusion, the fundamental truth of the spinel shines through – these are gems too big, too red, too eye-catching to miss. Like moths to a flame, they have attracted the powerful to make use of them and the curious to try and make sense of them. The spinel may have been known by many names throughout its history and endured a rollercoaster ride of reputation, but its essential qualities have never really been extinguished. As several of the most famous crowns ever worn attest, when it comes to making a statement in stone, spinel is simply a gem like no other.

7

Quartz

The Commoner Gem

*'The gem, if rarer, were a precious prize,
But now too common it neglected lies . . .'*

MARBODUS OF RENNES (1035–1123),
TRANSLATION C. W. KING (1870)

'CAN I TRY it on?'

I could hardly believe my eyes, or contain my excitement. In the elegant home of a collector, with record-breaking Warhols and contemporary masterpieces on every wall, I should have realized that the vault would contain nothing less than world-class jewellery. But I had not expected to find what was laid out in front of me. There was no warning of what was to come, just the anonymous box clicked open and its startling contents silently revealed. The moment I saw the famous baubles of blue chalcedony, I knew what I had been brought here to see: a set that had been made for Wallis Simpson, the Duchess of Windsor, wife of Edward VIII. I was looking at one of the most recognizable jewels of the twentieth century, crafted by one of its supreme designers, and fashioned for one of its most notorious style icons. It was not enough just to see the crown-shaped bracelets, their milky blue spheres topped off with sapphire and diamond accents, and the matching necklace with its floral centrepiece of huge petals of pastel blue chalcedony completed by a deep blue burst of sapphire. I also wanted, immediately, to wear them.

It was one of those special moments in the jewellery trade, that rare feeling of holding perfection in your hand or glimpsing its physical form up close. When I started in the jewellery business I promised myself that the moment I lost my ability to be delighted by the beauty of gemstones would be the moment I left it. I have expressly held on to the raw excitement that the very best pieces provoke, a reminder that the greatest jewels should always inspire emotion and evoke passion.

The blue chalcedony that had stirred such strong feelings in me was an example of the contradiction that is quartz: a gem group that is formed from one of the earth's most abundant materials, but which has also been crafted into some of the world's most valuable and distinctive jewels. Blue chalcedony is just one variety of a gem that exists in a dizzying array of different colours and forms – from black onyx to colourless rock crystal, lemony yellow citrine to purple amethyst and pastel pink rose quartz, carnelian in its multiple shades of red and orange, simple brown sard, translucent chalcedonies in white, brown and blue, and banded agates and sardonyx that could combine a mix of colours.

There are many more members besides of a family that stretches right across the colour spectrum, and which is found in relative abundance all over the world. Quartz should be a material of little value: as silicon dioxide, it is made from the two materials that together comprise almost three-quarters of the earth's crust. It is not a singular gemstone but a sprawling family of crystals stemming from a common source of one of the most plentiful compounds in existence. It can form almost anywhere, in multiple ways and incorporating numerous additional ingredients. Far from cheapening quartz, this ubiquity has been one of the secrets of its ageless appeal.

In the ancient world, quartzes were some of the first-found gem materials, better known and more widely discovered than many of their later-known alternatives. Archaeological discoveries show that carnelian was being fashioned into beads for jewellery as early as 4500 BC.[1] Whereas ancient civilizations might only have found gems like sapphire and emerald in small crystals, they were able to access quartz both in greater quantity and larger sizes. Jewellery could be furnished extravagantly and exactingly detailed designs created. Forms could be fashioned, elaborate images engraved, and whole scenes laid out: it was a medium for miniature sculpture in precious gems.

In addition, quartz's extensive colour palette made it appropriate for a wide range of symbolic uses, while its hard-wearing properties were ideal for carving and artistic uses beyond accent and decoration. Quartzes were available, they were colourful, and they were a craftsman's best friend. From the earliest days of the jewellery trade, these characteristics helped make them into some of the most important gem materials for the ancient world's most prominent civilizations – a striking signifier of wealth and power from ancient Egypt to the Roman Empire.

One early exponent was Queen Puabi, whose cylinder seal (found buried with her) suggests she may have been a female ruler of the Sumerians in Mesopotamia in her own right, during the First Dynasty of Ur that dates to the twenty-sixth and twenty-fifth centuries BC. Her magnificent burial costume features extensive use of carnelian and agate as part of an ornate assemblage including a crown of golden leaves and flowers, multiple necklaces, and a magnificent cloak formed of dozens of strings of beads reaching from her shoulders down to her waist.[2]

Around a millennium later, carnelian was also being used as part of royal jewellery that would become history's most famous discovery: the ornaments of Tutankhamun. His burial chamber, and the myriad treasures it contained, provides a window into the esteem in which quartz was held and the symbolic value it carried in the ancient world. Among the items discovered there in November 1922 by the archaeologist Howard Carter were several pectoral ornaments, designed to be worn protectively across the chest of the dead in their journey into the next life. An extraordinary example is the scarab pectoral, a piece as stuffed with symbolism as it is jam-packed with gems.[3] In its centre is the winged scarab,* crowned with the all-seeing Udjat Eye of Horus, flanked by raised cobras which carry the sun on their heads, and fringed at the base with a line of lotus flowers. Both the scarab and the lotus were symbols of the sun, also representing rebirth of the dead, while the Udjat Eye and Uraeus were markers for protection and royalty. This feast of solar, royal and funerary symbolism also conveys meaning in its use of gems: throughout the piece, the classic Egyptian combination of carnelian, lapis lazuli and turquoise forms the primary colour scheme. These locally – and regionally – sourced gems, a stunning visual counterpart to gold, were a symbolic grouping: the bright red of the sun, the deep blue of the heavens and the ambiguous blue-green that represented rebirth. This was a classic combination for the

---

* Uniquely, this particular scarab – a light translucent yellowish-green – was carved from Libyan desert glass: an amorphous (non-crystalline) silicate formed from sand melted by a high-pressure meteorite impact. It was probably not lost on the Egyptians that this stone really was out of this world.

Egyptians, also featuring on Tutankhamun's death mask and many more pectoral ornaments, demonstrating how the Egyptians used red carnelian as one of their 'Magic Triad' of symbolic stones. Another example of a significant quartz find from Tutankhamun's tomb was a scarab bracelet decorated with an amethyst, one of the earliest-known uses of this transparent purple quartz, and the beginning of its enduring role as a sign of royalty, rulership and riches.*

The Egyptians were far from alone among ancient civilizations in embracing quartz as a gemstone of choice. Continuing the tradition of Greek gem engraving, the Romans in particular made use of quartz's qualities as a canvas for carving, creating ambitious and extraordinary artworks from it. Their lapidaries took the art of glyptics to its apogee, both in intaglio (where the device was cut down into the gem) and in cameo (where the image or scene was carved in relief). Sardonyx – the brown- and white-banded variety of chalcedony – was especially apt for cameo work, if the artist was skilled enough to foresee the anomalies of the natural material and create a 3D tableau by carving correctly through its layers. A master gem engraver could work with as many as seven different variations of colour layers, alternating from white to dark by cutting the stone to different depths.

At the height of this genre, during the age of Augustus (27 BC – AD 14) and his heirs, the greatest artists of the era were commissioned to create detailed scenes for the leaders of the empire. Their notable victories in war, divine associations, and intended lines of succession were carved in relief from the stone, the figures rising up from its layers of smoky brown and milky

---

* This association has been wide-ranging and long-lasting. Catholic bishops still wear ecclesiastical rings of amethyst as an emblem of their authority.

white. This was a jewellery craft that utilized to full potential the qualities of the material: its attractive colours, its hard-wearing properties, and its availability as a broad canvas for this painstaking work. Large-scale cameos were important expressions of wealth and power, not least because of the cost of commissioning a lapidary to perform such complex and precarious work, where one mistake – a single chip or, worse, a break – would mean having to start all over again. At the other end of the scale, simpler cameo brooches, rings and earrings in sardonyx were so prolific in some regions in the third century that they appear to have been mass-produced for the wealthy middle classes.[4]

Some of the most famous examples of these Roman power pieces were commissioned by the emperors themselves, notably the *Gemma Augustea*, a lavish two-layered cameo that depicted the first Roman Emperor, Augustus, alongside a range of symbolic figures: from his two immediate heirs, Tiberius and Germanicus, to the goddess Roma, personification of Rome and protector of the empire, and other sacred figures representing the earth, oceans and civilization. Depicted in bare-chested, godlike form, with the *corona civica* (an oak-leaf crown granted to those who had saved the lives of fellow citizens) being placed onto his head, Augustus appears as a figure of supreme earthly power and one with close divine connections.[5] It is an outstanding example of how quartz cameos were used as propaganda pieces in high Roman society, the sheer size and layering of the material allowing lapidaries to carve scenes packed with power play and allegory for those at the highest levels of society.

Augustus also featured on an even larger piece in the same tradition, which was created several decades later. The *Grand Camée de France*, which was most likely carved either in the reign of Tiberius or his successor-but-one Claudius, depicts Augustus in deified form, one of twenty-three figures who appear on this enormous sardonyx slab, including four other past and future Roman rulers.[6] This cameo was to enjoy an even more decorated history than the one it portrayed: owned variously by Byzantine emperors, French kings and medieval popes, and ultimately by Louis XVI, who stashed it in what is now the National Library of France to avoid it being looted during the French Revolution.

The quartz carving tradition so embraced by the Romans had practical as well as decorative uses. Similar techniques were used to produce intaglios for signet rings where images were engraved into stones such as jasper, carnelian and amethyst. They also produced remarkable *objets* including the *Tazza Farnese*, a bowl of sardonyx with reliefs carved on both its interior and exterior and a true tour de force of ancient art. This much-discussed object has experienced an extraordinary journey through cultures and civilizations, prized by emperors and puzzled over by scholars who continue to dispute the purpose for which it was created. Likely originating at either the court of Cleopatra or the Emperor Augustus, it was subsequently owned by the Vatican and then sold to Lorenzo de Medici in 1471, before eventually ending up in the ownership of the Farnese family, another noble house of Renaissance Italy.

The *Tazza's* origins have been widely debated.[7] It is thought by some to have originated in Ptolemaic Egypt, as early as the third century BC, the stone sourced from as far afield as India or Bulgaria. Others have dated it to the first century and the *Pax Romana* that began with the reign of Augustus in 27 BC – bringing an end to the civil war that had followed the death of Julius

Caesar and the fall of the Roman Republic, at which point Rome's trading networks became steadily more prevalent. The debate over provenance has led to disagreement over the identity of the seven deity figures depicted on the face of the bowl: an Egyptian planetary allegory, or a Roman celebration of imperial power in the reign of Augustus. Less ambiguous is the gorgon figure carved out of its base, the familiar Medusa symbol that featured widely in Roman jewellery and was used to ward off malevolent spirits by averting the Evil Eye. This device is best known today as the face of the fashion house Versace.

The *Tazza*, one of the most enduring quartz creations, brings to life the paradoxical nature of the stone: as practical as it is beautiful, a material that demands both to be admired from a distance and to be worked on at close quarters. It is gemmology's blank canvas: where other stones are made into jewels, quartzes can be crafted into fully fledged works of art. The objects that result often have a functional as well as aesthetic purpose, and the *Tazza* was likely no exception. Scholars have speculated about its potential uses, with one suggestion being that it would have played a role in Roman religious ceremonies, as a container for wine to be spilled in a form of symbolic sacrifice, the downward face of Medusa on the bowl's exterior serving to ward off unwanted presences.

That suggestion carries an interesting echo of something we know for certain about other quartz varieties: the association of amethyst with the grape and vine, and ancient ideas about alcohol and drunkenness. The word 'amethyst' derives directly from the Greek ἀ-μέθυστος (not drunken). In his *Natural History*, Pliny the Elder explained that it was said that the name was derived from 'it closely approaching the colour of wine', while simultaneously distancing himself from the extended idea: 'charlatans falsely claim that these stones prevent drunkenness, and that this is how they got their name'.[8]

This is just one of the many symbolic associations that have grown up around quartz, hardly unusual for a gemstone, but especially plentiful because of the range of colours in which it is found. Throughout history, the different varieties and variations of quartz have proved to be the perfect physical medium for the reflection of the human psyche: a one-stop shop for a myriad of belief systems and symbolism.

A common example is banded agate or sardonyx, valued in antiquity for its protective powers for warding off the 'Evil Eye', and still used today for jewellery intended to safeguard its wearer from harm. The heliotrope (or bloodstone), a dark-green gem with blood-red spots that reveal its iron-oxide impurities, is another quartz that has taken on a symbolic life of its own. It was said that Roman gladiators and soldiers wore them as amulets for the arena and battlefield, and also considered them medicinal, as an aid to circulation and a treatment for blood and bowel disorders.[9] In the medieval period, the flecks of red associated bloodstone with Christian sacrifice, making it a popular choice for depictions of saints and the Crucifixion.[10]

Such symbolism is everywhere in quartz's extensive canon. In its various incarnations, quartz has been widely imbued with power, protection and promise, and perhaps never more so than today. The discovery of plentiful deposits of Brazilian quartzes in the nineteenth century – including amethyst and the ever-popular rose quartz – has been large enough to supply and fuel an entire industry of New Age crystal healing: a modern belief system with roots reaching firmly back to antiquity. Once again, gems show us the threads of connection between civilizations and cultures spanning millennia of human history.

THE WIDE-RANGING uses of quartz reflect a gem family that can emerge from the earth almost anywhere in the world. Quartzes grow in different underground locations and various kinds of rock; many result from the magmatic or hydro-thermal flows caused by underground volcanic eruptions, with the gem material forming in pockets as the melt solidifies. This creates crystal-lined geodes, hollow rock forms whose insides, once cracked open, seem to be bursting with crystals. The upshot of these various environments, with crystallization occurring at different speeds and encompassing a variety of different min-erals picked up along the way, is that quartz emerges as a common compound manifested in numerous forms. A teeming family of crystals can often appear to bear little relation to one another – some with deep, rich hues while others are pale pastels; some transparent while others are opaque, some with their crystals visible to the naked eye and others where they cannot be seen at all.

It is this final distinction that gemmologists typically use to categorize quartzes. First you have the macrocrystalline quartzes: amethyst, citrine, colourless rock crystal and pale pink rose quartz being just a few of them. These are the show-off variety of quartz, where you can see the individual crystal faces of trans-parent gem material. Then there are the microcrystalline quartzes, where those crystals exist in the structure, but on a much smaller scale; they are essentially hidden within (hence they are sometimes also known as cryptocrystalline), creating a more translucent or opaque appearance. Microcrystalline stones such as agate, chalcedony and carnelian are the workhorse members of the family, thanks to their tightly knitted crystalline structure. An aggregate microcrystalline variety, jasper, takes this structure to its extreme, appearing more like a rock than a traditional gem material. This creates an effect a little like a well-built Lego model, with the constituent parts interlocking tightly but with no planes of weakness, forming a gem material that is notably durable. It is this property that has made these stones the best friend of lapidaries through history. They respond beautifully to the carver's drill, blade, wheel and saw, forming the basis for some of the most elaborate pieces of gemmological art ever made.

The admiration for quartz throughout human history can be explained by the three categories on which all gemstones are ultimately judged: beauty, durability and rarity. Quartzes in their many colours are beautiful, or at least they have been thought so by the numerous civilizations that have used them in decorative and symbolic jewellery. They are durable, especially the microcrystalline varieties that are such a lapidary's dream. And although quartzes are no longer in actual fact rare, to the ancients they would have seemed so, and at various points in history they actually have been. According to Pliny the Elder, writing in the first century AD, the variegated quartzes 'were once held in high esteem, but now have none', and the brown sard was simply 'common': both commentaries on their more recent availability in his day.[11] In the case of amethyst, rarity was a reality until the early nineteenth century. Before this, the deep purple quartz from Siberia was prized on a par with emerald, ruby, sapphire and even diamond.[12]

Quartzes are in many ways the archetypal gemstone: objects that have variously been venerated for their beauty, harnessed for their value and practicality, utilized for symbolic and political purposes, and which have formed the basis for a flourishing set of mythologies. The intersection they provide between the aesthetic and the functional has made them as enticing to the society trendsetters of the modern world as they were to the rulers and power brokers of civilizations thousands of years before.

IN APRIL 1987, just over half a century after the abdication of Edward VIII that shook the monarchy to its core, and ten months since the death of his widow Wallis Simpson, the Duchess of Windsor, her jewellery collection was auctioned. Sotheby's pitched what had been a circus tent by the bank of Lake Geneva for the thousand bidders who attended in person, with hundreds more participating remotely. Over two days some of the most iconic pieces of the twentieth century were sold, including the Cartier diamond and onyx panther bracelet that had been made for her in 1952 – which would become the most expensive item of its kind ever sold at auction – and a decadent flamingo brooch, its tail feathers bedecked in emeralds, rubies and sapphires.[13] Bidders included Elizabeth Taylor – whose own collection would later break the record set for a jewellery collection by the 1987 sale – working through a broker and disguising her identity under the initials JB. She bought the diamond Prince of Wales brooch in memory of her former husband Richard Burton, and the friendship the famous couple had formed with the duke and duchess, one of few marriages to know the same notoriety. Elizabeth Taylor later said it was the only piece of jewellery she had ever bought for herself.[14]

In total, approximately $50 million was raised through the sale, more than six times the original estimate. The 'sale of the century' was part of the institutional memory of Sotheby's, one that the people I worked with still talked about fifteen years later. But amid the pomp, glamour and nostalgia, some things had been almost lost and forgotten. The name of Suzanne Belperron, one of the duchess's favourite jewellery designers, who had herself died four years earlier, featured only fleetingly in the catalogue. Of the fifteen pieces in the collection now identified as Belperron's, just five were attributed to her, and then only tentatively, including the blue chalcedony suite.[15] At the premier jewellery event of the century, one of its most iconic designers had receded almost into obscurity.

This then-forgotten trendsetter had lived an extraordinary life, one that began at the dawn of the century and would be interwoven with some of its worst history.[16] The daughter of a baker, Suzanne Belperron came to Paris in 1919, the First World War having ended, and the Art Deco era about to begin. She launched her design career at the reputed Maison Boivin and formed her most significant alliance after leaving it in 1932 to work with the Jewish jewellery dealer Bernard Herz. It was in the 1930s, alongside Herz, that her reputation began to bloom. *Vogue* hailed 'a revival in the art of jewelry' when the fashion designer Elsa Schiaparelli appeared in its pages wearing a Belperron necklace in 1933.[17] Another article the following year described her work as 'new and barbaric'. The acclaim and open mouths had been well earned. Art Deco was the prevailing wind in fashion, which for jewellery meant angular, geometric, streamlined and often monochrome. This was a template Belperron happily ripped up. Where convention dictated minimalist colours, she went for bold, clashing tones: reds and purples, blue on lilac, pioneering colour combinations that would not become mainstream until half a century later. Where the consensus preferred precious stones, she readily mixed them in with other less fashionable quartzes – as well as chalcedony, she often used amethyst, smoky quartz and rock crystal. And where the Art Deco preference was for clean lines and precision, she luxuriated in huge shapes and complex, challenging-to-craft motifs from nature. She was less Belle Epoque, more *Sex and the City*: not so much a breath of fresh air as a hurricane blowing through the ateliers of inter-war Paris.

A feminist force of nature, she was the perfect creative counterpart for another woman with willpower to spare – Mrs Simpson, as she was until she became history's most famous double divorcee in 1936. The nonconformist designer and the woman who had sparked a constitutional crisis made an apt pairing: a meeting of minds both in style and attitude. Wallis was no more afraid of standing out than Suzanne was of breaking from tradition. 'I am not a beautiful woman. I'm nothing to look at, so the only thing I can do is dress better than anyone else,' Wallis once stated.[18] Less charitably, the writer James Pope-Hennessy commented that, 'She was flat and angular and could have been designed for a medieval playing card.'[19] Wallis knew that blending in was never an option, either socially or aesthetically. So she embraced the extremes of fashion, choosing to shock in such designs as the Schiaparelli 'lobster dress' designed by the surrealist Salvador Dalí and which she was photographed wearing by Cecil Beaton for *Vogue* in 1937. For this unconventional style icon, Belperron's outlandish designs and striking motifs were a natural fit.

The pair were connected by more than fashion. Both were women before their time – one producing designs that were decades ahead of mainstream trends, the other falling in love with a future king in an age when it was inconceivable that a monarch might marry a divorced woman and keep his crown. Both were known to be challenging. Belperron, whose intricate designs would often exasperate her craftsmen, was famed for her customarily curt response to their protests: '*Débrouillez-vous!*' ('Sort it out!')[20] Nor was the duchess a wallflower when it came to the business of fashion. 'She was easier on design than she was on price, and she was a terrible stickler on design,' recalled an executive at Harry Winston, her New York jeweller, at the time of the 1987 sale.[21] Both women also had lives in which a great love loomed large, Suzanne's eminently more tragic than Wallis's. Suzanne's multi-faceted relationship with Bernard Herz was brought to a shattering end by the Nazi occupation of France in 1940. Despite her efforts to protect him and the firm's Jewish clientele – reputedly swallowing pages from their record book while under arrest – Bernard was first detained at the Drancy concentration camp north of Paris, and later Auschwitz, where he was killed in 1943. She continued to operate the business, which had been placed in her name to protect it, and later returned it to Herz's son Jean, after he returned from his own internment at the end of the war. The two would go on to run the relaunched Herz-Belperron company together until her retirement in 1974.[22]

By Belperron's death in 1983, she had largely faded from view, not least because of her dogged refusal ever to sign her work. 'My style is my signature', her most famous saying went. And for a time, this doctrine threatened to consign her to obscurity. Yet the splash made by the Duchess of Windsor sale, and subsequently the work done to recover and research her archives, helped restore her deserved status as a pioneer of style. In the years since, Belperron has regained her repute as one of the legends of jewellery design and a true twentieth-century trailblazer: 'modern, before the world was', in the words of *The New York Times*.[23] Her motto has become the statement of enduring legacy she had intended: that you can look at a piece and know immediately that it is Belperron. When her blue chalcedony set was laid out in front of me, I recognized it like an old friend. But I could not have anticipated the feelings that handling this remarkable object, a true queen of quartzes, would provoke. It would be a memorable encounter with jewellery that was at once among the most powerful and the most elegant pieces I had ever handled, evoking the spirit of the two extraordinary women who had respectively brought it into being, and made it into an icon.

TO UNDERSTAND fully a piece of jewellery that was made to be worn, you really have to try it on. And once I had strapped the double cuff bracelets to my wrists and fitted the clasp of the necklace, I could immediately feel that these were pieces heavy with meaning, dripping not just with wealth but with power. It was like I imagine a suit of armour would feel – as if I were dressed not in bracelets and a necklace, but gauntlets and a breastplate. To wear these jewels was almost to step into the world of Wallis and Suzanne, a social icon so infamous and a creative genius so luminous that they could dare relegate diamonds and sapphires to incidental accents. Instead, they made their statement in the ambiguous form of chalcedony, a material that would have carried only a fraction of the same worth without the meaning they instilled in it.

These were also pieces that illuminated the peculiar qualities of quartz, a family of stones that is packed with paradoxes as enticing as its many colours and forms. Here, it was impossible not to enjoy the implicit contrast between the feminine and the masculine: the marriage of bulky shapes with subtle sheens of colour, the seemingly huge baubles with their elegant garnish of graded blue hues of sapphire, and the delicate craftwork of the floral centrepiece set against the brash shapes surrounding it. Somehow, Belperron's genius had created a piece that was simultaneously subtle and strong, delicate and dramatic, common and courtly. And of course the contrast was completed by the spare, beyond slim figure of the woman these sizeable pieces were designed for, resplendent in her favourite blue.*

The necklace and bracelets – which the duchess also wore with a pair of matching earrings, initially designed as the original necklace clasp but replaced early on with the more striking flower motif – were replete with symbolism as much as they were packed with flourishes of design. When the bracelets are laid flat, the baubles topping the semi-circular bands, it becomes clear that they are crowns: they were in fact known as '*couronnes*'. The then Prince of Wales is thought to have commissioned the piece around 1935, the year before his short-lived ascent to the throne, by which time the events that would culminate in his abdication were already in motion. The future Edward VIII had met Wallis Simpson in 1931, introduced by his then mistress Thelma Furness. By 1934 they had begun a romance in which jewellery would play a central role, intrinsic to this controversial couple's language of love. The future king showered his intended queen with jewels which, like Prince Albert before him, he took a personal role in creating, working with the Paris workshops of Cartier, Van Cleef & Arpels and Herz-Belperron on gems that

---

* The Duchess of Windsor may have been a dainty figure, but she did not always treat her jewellery delicately. The chalcedony suite was broken on more than one occasion and had to be sent back to Belperron for repair.

had been gifted to him and consulting on designs. Jean Herz recalled an occasion on which he had seen one of the most powerful men in the world on his hands and knees in a suite in Paris's Le Meurice, poring over a selection of Belperron designs that had been laid out on the carpet.[24] These were exquisite and above all expensive gifts: royally commissioned and with price tags to match. A Van Cleef bill from the year of the abdication shows that the king paid £16,000 for a series of platinum, ruby and diamond pieces; at any time since the turn of the twenty-first century, the same shopping trip would have cost him comfortably north of half a million pounds.[25]

As important as the value and beauty of these jewels was their hidden meaning. Many of the pieces that passed between Edward and Wallis carried inscriptions. 'WE are ours now', read the emerald engagement ring he gave her – a customary self-reference to the couple by their initials. As well as his love and commitment, jewels also expressed Edward's frustration at his perceived predicament. One of her favourite pieces was a bracelet of crucifixes that were added at different times, each bearing its own meaningful date and inscription – a piece I had worked on early in my career. One of these, dating to March 1936, early in his abortive 325-day reign, read, 'The Kings Cross God bless WE', almost certainly articulating his frustration at the limiting obligations of royal duty – his cross to bear, and one he had resolved to lay aside.[26]

The chalcedony suite, therefore, must be seen within the context of a relationship in which jewellery was not just a preferred gift but a primary source of symbolic self-expression for a couple who, in their formative years, adopted secrecy as a necessity. Against that backdrop, the meaning of the *couronne* bracelets is not hard to deduce. They were hidden-in-plain-sight crowns for the woman this uncrowned king wished to be his queen, but knew never could be. His apparent obsession with building her extraordinary jewellery collection may be seen as an attempt to replace and remake the Crown Jewels that he knew would never be available to the woman he loved. This becomes even more poignant in the context of a later event that rankled with the couple in exile – the refusal to grant the now Duchess of Windsor the title of Her Royal Highness, which would customarily have been hers by marriage to him. A couple that believed itself to have been vindictively exiled and excluded from the Royal Family had only their own symbolism and stylings to fall back on. Wallis's 'uncrowned jewels', as a BBC documentary at the time of the 1987 sale characterized them, were perhaps the most important part of this.

It can hardly have been accidental that as daring a designer as Belperron chose to use quartz, one of the commonest gems, to craft a statement piece for this alternative set of crown jewels. The choice of blue chalcedony, stylishly surprising and impressively impactful, made a statement both about the woman who had designed it and the client it was for. And, crucially, it offered utility. The magnificent, gaudily eye-grabbing clasp is the kind of piece that only the massive and easily carved chalcedony could enable – its availability and durability enabling a boldness of design that would have been inconceivable with more precious stones. For Belperron's style of mixing precious stones with those then deemed 'semi-precious', in a way many contemporaries would have regarded as inconceivable, this was perhaps the signature piece. It celebrated her own genre-defining creativity, her client's ambiguous quasi-royalty, and the practical properties of the quiet marvel that is quartz.

8

Diamond

The King of Gems

'The most highly valued human possession, not only among gemstones, is the diamond, known for a long time only to kings, and then only very few of them.'

PLINY, FIRST CENTURY

I WAS about to descend deep into the belly of the most famous diamond mine in the world. I was en route to Namibia and Botswana – two of the world's biggest diamond-producing countries – with a layover in South Africa on the way. The town of Cullinan was only 100 kilometres from Johannesburg, and the Cullinan Mine – once known as the Premier Mine – had produced some of the most important diamonds in history. It was springtime, and surprisingly picturesque: the bleak surface operations stood in stark contrast to the surrounding avenues lined with violet jacaranda trees exploding in full blossom. As I was chatting to the mine manager, he pointed to a helicopter flying overhead and explained that an important stone had just been found and was on its way out. The next moment our guide – a retired miner who had stayed on to take tours – arrived and we were on our way in. He entered the dark like it was his second home.

Kitted out in overalls, hard hats, lamps, goggles and emergency rescue packs strapped to our backs, we were given a full safety demonstration before we could enter the metal cages which would take us down. The lift shaft that transported us held two dozen miners, each starting their next shift. No one spoke a word and it wasn't until the cage had emptied that I saw the sign and realized where we were: '763 Level', three-quarters of a kilometre below the ground where the world's largest diamond had come to light.

I stepped out of the cage and into an enormous cavern reaching further than the eye could see, its height able to accommodate industrial diggers and drills. Antique-looking mechanized trolleys, like something out of a fairy tale, were running along a miniature railway through the tunnels carrying ore to be dumped at various points on its journey. The clanking and whirring of heavy machinery faded in and out, together with intermittent dynamite blasts echoing through the spaces, throwing up regular, blinding dust clouds. This was a thoroughly modern, mechanized operation, a world away in every sense from the discovery that had ensured the mine's celebrity status, in what initially seemed to be a superficial find just over a century ago.

DIAMONDS FORM deeper in the earth than any other gemstone, yet this one was extracted with nothing more than a penknife. The humble implement belonged to Fred Wells, the son of an English bootmaker who had lived most of his life in South Africa and was one of the managers of the Premier Mine in Gauteng. The mine had been open for less than three years when he made his stunning discovery, late in the working day on 26 January 1905. So large was the crystal that Wells could see peeking out from the open pit wall of No. 2 Workings that it broke the blade of his pocket knife as he tried to pry it out of its rock. He did not initially think it could be a diamond, believing that someone must have 'planted this huge chunk of glass' for him to find as a practical joke.[1] His scepticism was shared by his boss, the Premier's owner Sir Thomas Cullinan, who received news of the find by telegram that evening, and supposedly remarked to his dinner guests: 'I expect they are wrong. It is probably a large crystal.'[2]

Their doubts would soon be dispelled. Wells had found something unprecedented: a rough diamond that, at four inches across and over two and a half high, weighing in at 3,106 carats, was too large for anyone to close their fist around. Moreover, this monster was likely a fragment, thought to be less than half of the original crystal. A find that had begun with suspicion would quickly become a phenomenon: when it went on display a week after Wells had extracted it, at a bank in Johannesburg, 9,000 of the city's residents queued up to see it.

For Sir Thomas, the diamond that would bear his name was not as obvious a prize as it might have seemed. Its novelty value was profound, but its market value less so: the previous record holder, the 995-carat Excelsior, had gone unsold for twelve years after being discovered. After being sent to the Premier's sales agent in London – using simple recorded postage, while a decoy was shipped under ostentatious guard – the Cullinan similarly endured two years of being shown to potential buyers without being snapped up. The uncut diamond might have been a remarkable rock but it was not, yet, a remarkable sight. 'I should have kicked it aside as a lump of glass if I had seen it in the road,' commented its eventual owner, King Edward VII, after the Cullinan was brought to Buckingham Palace for a private viewing. It was clear that the mega diamond would need to be cut to bring out its brilliant best. But this work would be difficult, not to mention expensive, and for two years, the diamond languished in a bank safe, unsold, and uncut.[3]

The impasse was finally broken by the Prime Minister of Transvaal, General Louis Botha, who in 1907 tabled a parliamentary motion for the province to purchase the diamond as a gift for King Edward VII, the ruler of what still was, until its independence three years later, a British colony. The price was £150,000 (£15–20 million today), 60 per cent of which was written off by the Transvaal government as a tax deduction applied on diamond-mining proceeds. The King was not initially keen to accept such a gift, but was convinced to do so by Winston Churchill, then Under-Secretary of State for the Colonies. For his part in the deal, the Transvaal government sent Churchill a copy of the diamond, which he would show off on a silver salver when guests came to visit. So unimpressive was the rough replica that on one occasion the guest in question thought she was being offered a badly strained jelly, and, with a dismissive glance, simply replied, 'No thank you.'[4]

Once the diamond had been presented, on Edward's sixty-sixth birthday, the question quickly became how the gargantuan stone could be turned into something more presentable. The King appointed Europe's foremost experts in the field, the Asscher Diamond Company in Amsterdam, to take on the mighty task. Brothers Joseph and Abraham, grandsons of the founder, had built a continental reputation as diamond cutters. In 1902 they had achieved the first ever patent for a diamond cut, a square, 58-facet model with clipped corners.* The following year they

---

* Known as the Asscher cut, it was a favourite in Art Deco jewellery for its clean lines, and its ability to reflect light internally like a Hall of Mirrors.

were commissioned to cut the Excelsior, the Cullinan's predecessor as the world's largest diamond, into ten separate stones. They were the obvious choice for the most high-stakes diamond-cutting operation ever attempted.

No part of this mammoth assignment would be simple. Even the stone's transport from London to Amsterdam entailed risk: the diamond was carried by Abraham Asscher in his coat pocket, while an empty box sailed in parallel under the guard of the Royal Navy, a decoy mirroring the stone's first journey from the Transvaal. Once it was safely in the Asscher workshop, the cutting could begin, but it did not all go according to plan. After weeks of careful preparation, with a 2-centimetre notch having been sawn into the diamond, the moment finally came for Joseph to attempt the initial cleaving.* Yet when he placed the steel knife in the groove and struck it with a hammer, in a flashback to its original extraction, it was not the diamond that broke into pieces but the knife.†

A second attempt proved more successful, separating the Cullinan into two pieces that would then be cut into nine principal stones, plus a wealth of satellites. Although the initial action had been the nerve-wracking part, in many ways the real work followed: eight months of long days to cut and polish the products of the Cullinan, the two largest of which would become part of the Crown Jewels: the Great Star of Africa (Cullinan I) that adorns the Sovereign's Sceptre, and the Second Star of Africa

---

* The preliminary separation of a large rough into the pieces that will then be worked on individually, splitting the stone along its natural planes of weakness. A critical and high-stakes stage of the process in which the stone is at risk of shattering if pressure is misapplied.

† Joseph is sometimes said to have fainted upon first striking the Cullinan, an unnecessary embellishment to an already nail-biting event.

(Cullinan II) that is mounted in the Imperial State Crown. Given their size, each was cut with even more facets than the Asscher signature: seventy-five and sixty-six respectively. These went to the King, and the brothers retained the remainder to sell as their fee.[5]

As a massive slice of diamond history, the Cullinan is notable not just for its unprecedented size – a record that has remained ever since – but for its story, one which encompasses much of what has made the diamond so mysterious and revered. It showed that there is nothing like a famous diamond to create intrigue and subterfuge, as it travelled around the world courtesy of a series of misdirections, with constant fear that attempts would be made to steal it. It demonstrated the importance of the cut, and how the crystalline carbon can be brought to life by a manufacturing process that brings out its extraordinary ability to reflect, refract and disperse light. And it underlined that the diamond is above all a political gem, one that has been a source of great wealth, exploitation and power exchange throughout history. In the case of the Cullinan, the largest diamond ever discovered soon became a diplomatic token. When General Botha's Transvaal government presented it to Edward VII, their aim was to build bridges after the Boer War, to soften the request of a £5 million loan (£500 million today) from the British Government and to pave the way for the independence that they would soon achieve. 'In many ways, it was the original peace diamond,' according to Mark Cullinan, great-grandson of the original mine owner, Sir Thomas.[6] Contrary to the negative reputation associated with diamonds in later years, the Cullinan had become an expression of reconciliation. It would heal the wounds of battle and lay the path to a new future.

HOW DIAMONDS like the Cullinan came close enough to the earth's surface for humans to discover and extract is the product of two geological marvels. The first is the formation of the crystal itself, the compression of carbon atoms under conditions of sufficient heat (approximately 1,200 degrees Celsius) and pressure (more than 40,000 atmospheres) that they combine into structures of immense strength – able to survive almost anything that nature can throw at them.[7] The diamond's innate hardness is deep-rooted: whereas almost all other gemstones form within the earth's crust, the diamond's origin is way down in the mantle, several hundred kilometres below the surface.

The formation of the diamond is just the beginning of its high-octane journey. The next leg follows courtesy of volcanic eruptions that begin deep underground, where the melting of the mantle creates a magma rich in carbon dioxide, one that explodes upwards like the contents of a shaken bottle of fizz. This underground jet gains impetus as it goes, with xenoliths (literally, foreign rocks) including silicates and diamonds joining the mix. These mineral additions are the geological equivalent of the effect of dropping Mentos into a bottle of Diet Coke – causing the release of carbon dioxide, making the magmatic mixture more acidic and ultimately helping it foam and explode upwards, towards and through the surface. The magnitude of such volcanic events can only be estimated, since most eruptions of this kind happened between 70 and 150 million years ago.[8] The diamonds they carried were even older: between 1 and 3 billion years old.*⁹ Within the context of the history of the earth itself – approximately 4.6 billion years old – diamonds are ancient.

The result, once cooled, is a carrot-shaped intrusion of a rock known as kimberlite, after Kimberley in South Africa's Northern Cape, where primary diamond deposits of this kind were first discovered in 1871.† These kimberlite 'pipes' are the natural world's delivery system for diamonds. Only a small percentage of pipes will contain any meaningful quantity of diamonds and only a fraction of those are commercially exploitable.¹⁰ The modern method of extracting them entails the heavy industrial blasting that I observed at the Cullinan Mine, followed by the crushing and processing of kimberlite ore, breaking it down into smaller and smaller pieces so as to discover whether diamonds are hiding within. Yet for most of the diamond's history its geological origin was unknown and discoveries were of stones that had been eroded and washed away, mined alluvially from extant or long-buried riverbeds. Some of those diamonds were washed into the ocean, where they are recovered today by specially designed 'crawler' ships that dredge and sift gravel from the seabed.

Diamonds are formed under conditions of extreme heat and pressure, and they are brought into our realm by volcanic eruptions with no parallel in the earth's modern history. It should be no surprise that the resulting stone – the one that has survived all this – is like no other, with physical and chemical qualities that explain its place at the peak of the precious gem hierarchy.

---

* It was not until the 1980s that geologists were able to date the rocks with accuracy. Their tests confirmed that the diamonds were not formed within this host rock, only transported by it.

† Diamonds may also be transported in a similar process by lamproite, a volcanic rock rich in potassium and magnesium.

It is its hardness for which diamond is most famous. Sitting at the top of the Mohs scale (which ranks how comparatively resistant a mineral is to scratching), diamond scores a perfect 10, above ruby and sapphire at 9. Yet that understates the level of difference: in absolute terms, the diamond can be more than four times as hard as corundum, courtesy of its tight and symmetrical atomic structure.[11] This hardness – only a diamond can scratch another diamond – explains one secret of the diamond's beauty. It allows it to be polished to a level unlike any other, giving it supreme and adamantine lustre (the reflection of light off the surface). The polishing – a separate process to the cutting, performed by its own specialists – can be a major undertaking in its own right: for Cullinan I, it began in March 1908 and was not finished until September, with a team of expert polishers working fourteen-hour days to complete the job.[12]

Yet hardness is only part of the diamond's secret formula. What makes the stone truly visually remarkable is its high refractive index, the ability to slow down and bend the light waves entering it. If cut correctly, light entering the gem will bounce off multiple internal surfaces, break up into its component colour waves (the dispersion of light known as a diamond's 'fire'), and bounce back out into the eye of the onlooker, creating an internal reflection we call brilliance. This combination of reflection, refraction and dispersion is at play in all gemstones, but never better than with a diamond that has been cut and polished to perfection. There is no secret to the diamond's long-standing reputation as the King of Gems: the chemical and physical marvel of its creation gives it an appearance and a relationship to light that is both magical and unmatched.

It is the simplicity of the diamond that makes it so stunning – the perfection that arises from carbon alone, the atoms covalently bound together in a cubic structure, with perfect symmetry along all planes and axes. Yet carbon is not quite on its own within the structure of a diamond. Almost all natural diamonds – 95 per cent – contain traces of nitrogen in small clusters (Type Ia), which can lend the diamond a faint 'cape' yellow tinge.* A much smaller subset (Type Ib) have isolated nitrogen atoms which imbue the stones with a bright 'canary' yellow hue.

It is the minority without any impurities at all (Type IIa) that are the truly special diamonds, which make up almost all the world's most famous and highly valued examples, including the Cullinan.† Like the Cullinan, Type IIa diamonds are often completely colourless, the whitest of the white, with wonderful transparency. These are remarkable gems, with a purity and crystalline clarity that make you feel like you are looking into a perfect, limpid pool of water. Even more exceptional is the small percentage of these Type IIa diamonds with structural defects in the crystal lattice which create a pink colour. The final category (Type IIb) contain traces of boron, which can turn the diamond blue, and allow it to conduct electricity.

Like all gemstones, diamonds have origins that have developed as markers of quality in their own right. Today, India hardly features among the world's major diamond producers, contributing less than 0.1 per cent of the annual global yield. Yet until the eighteenth century it was not just the primary source but – with the exception of Borneo – the only source of diamonds in the world. It was the centre of a trade that stretched through time

---

* 'Cape' as a term for 'off-white' diamonds was developed after the discovery of South African deposits, which yielded so many of these faint and light-yellow stones that the term became generic.

† The Cullinans I and II have been graded as D colour (the highest grade for a colourless stone) and Potentially Flawless (meaning they almost have perfect clarity) on the standard industry classification system.

and around the globe, from the eastern tip of Hellenistic Greece in the third century BC to fourteenth-century Venice, the birthplace of Europe's diamond-cutting industry.

Once dominant, the Indian diamond mines now hardly produce, but they have retained their allure thanks to the large proportion of legendary Type IIa stones they yielded in their prime, known collectively as 'Golconda diamonds'. The Kingdom of Golconda, nestled between two major rivers flowing out to India's east coast, was home to a thriving diamond industry between the sixteenth and eighteenth centuries. Diamonds from the kingdom's mines would be brought into its capital (also Golconda), a fortress city on the western edge of modern Hyderabad, where they were cut, polished and traded. Until the discovery of Brazilian diamonds in the 1730s, Golconda was the only place in the world where diamonds were known to emerge with any reliability, making it a magnet for intrepid merchants and explorers, who in turn helped to build its global reputation. Even long after its heyday, and with manifold alternative sources of diamond well established, Golconda retains much of its lustre. Like many labels in the trade, the idea of the 'Golconda diamond' has expanded well beyond its original meaning, used to describe Type IIa diamonds not just from the area in question but sometimes other countries entirely.*

India is not just famous in diamond history for the quality of stones it has produced. Early Sanskrit literature makes clear how

---

* Geographical confusion is common with diamonds, which generally lack the diversity of mineral impurities and chemical variation used by gemmologists to identify the origin of a stone. If a diamond has been removed from its source and lacks origin certification, it can be impossible to tell for sure where in the world it came from.

long diamonds have held a central role in Indian culture and society: the *Arthashastra*, a treatise on statecraft, ruling and finance dating to the early third century BC, includes information on where diamonds were found (alluvially, in riverbeds and shallow mines), their colour ('like that of a cat's eye, the urine of a cow, or like that of any of the gems'), their qualities ('that which is big, heavy, hard, regular, capable of scratching on the surface of vessels, refractive of light and brilliant, is best'), how to measure them using grains of rice as a counterweight, and how merchants who traded them should be taxed.[13] More than 2,000 years before the diamond rush that began after stones were found in Kimberley in 1866, heralding the birth of the modern diamond industry, texts such as these underline the sophistication of the ancient world's diamond trade. The era in which diamonds have been scrutinized to assess their value, traded across borders and continents, and pursued to find the most brilliant examples, spans not centuries but millennia. The question of how to find, fashion and measure the perfect diamond gripped the merchants and scholars of the ancient world; and it is one that, even in light of all the advances in knowledge since, is no less animating to their successors today.

E VER SINCE diamonds have been pulled from the ground, stories have been told about them. One of the most persistent, which appears in different versions spanning multiple cultures and heroes including Marco Polo, Sinbad the Sailor and Alexander the Great, is that of the Valley of Snakes. Most versions of the tale include the same key components: a valley filled not only with priceless gems or diamonds is also teeming with serpents. The way to retrieve the treasure, our heroes are shown, is for pieces of meat to be thrown into this glittering snake-pit, whereupon eagles will swoop in and pick up not only their free lunch, but the jewels that have stuck to it. Once they have brought both back to their nests, the stones can be safely retrieved.

Like all great myths, there are some grains of truth stuck to this old story. As a hydrophobic mineral, diamond is indeed a very adhesive gem: it was common practice in the industrial diamond mines of the nineteenth and twentieth centuries for conveyor belts carrying crushed kimberlite ore to be extensively greased for this reason. Even with the use of X-ray sorting in today's mines, 'grease tables' still sometimes play a role in recovering the smallest and most elusive stones.

The second kernel of truth concerns the birds, which are very real actors in the contemporary diamond trade. In Alexander Bay, near South Africa's border with Namibia, on the Diamond Coast where the gems were first discovered in 1925, it has been illegal since the late 1990s to raise pigeons in the area, and any found by the authorities are condemned to be shot. As the journalist Matthew Gavin Frank recounted in his book *Flight of the Diamond Smugglers*, that edict resulted from the thriving trade in diamond smuggling in which miners – routinely searched on exit to ensure they were not taking their finds with them – had resorted to flying the gems out attached to the legs and bodies of homing pigeons. Frank met a thirteen-year-old mine worker – illegal child labour remains rife in the diamond industry – who would sneak his last remaining pigeon, Bartholemew, into the mine in his lunchbox, tying tiny bags made from cornmeal sacks under both legs and wings to fill with diamonds. This apparently charming avian version of *The Great Escape* does not always have a happy ending: as well as pigeons being shot, those that have touched down too soon due to carrying too much weight may land on a crowded street, and have their bodies literally torn wing from wing as people scrabble to retrieve their precious cargo.[14]

What links the snakes of ancient myth with the pigeons of modern mining is the perpetual possibility that, where diamonds are being excavated, they are also being stolen. The Valley of Snakes legend is only the most persistent of the narratives that have been used to deter would-be pilferers. In the *Arabian Nights* version of the story, the snakes can poison the diamonds by licking them: one origin of the old wives' tale designed to stop people from swallowing diamonds as a means of walking off with them. The stories may have dried up, but the excessive precautions against theft remain: in the 'red zone' of the Karowe diamond mine in Botswana, where the diamonds are sorted, they may only be touched using gloves built into the glass walls separating sorter from stones. In addition, every single person – including the company's chief executive – is strip-searched on departure each day.[15]

Such stories designed to repel theft are no less pervasive than those that have been crafted to ensure sales. No gemstone may have more combined qualities than the diamond, but equally none has had more effort put into marketing its virtues. Because diamonds enjoy such a close association with eternal love, as the go-to gem for an engagement ring, it is natural to assume this has always been the case.

It is true that there is a long history of diamonds being linked to the theme of enduring love. A seventeenth-century history of precious stones noted of the diamond: 'because of its hardnesse [it] can scarce be injured by the power of the fiercest fire; and for this cause it may be used symbolically as a signification of constancy.'[16] Earlier references to diamond rings as engagement tokens come from Europe in the fifteenth century, relating to notables including Maximilian I of Austria and the Italian aristocrat Costanzo Sforza.* The latter's marriage to Camilla d'Aragona in 1475 was celebrated with a poem redolent of associations that would become universal: 'Two wills, two hearts, two passions / Are bonded in marriage by a diamond.'[17]

Yet fragments of a long history and the obvious association cannot alone explain how diamonds came to dominate the market for engagement rings, to the point where an estimated three-quarters of brides in America walk down the aisle wearing one. This outsized cultural role was the product not of history's gradual accumulation, but the power of advertising and marketing: a deliberate – and brilliant – campaign to uplift the diamond, via 'the use of propaganda in various forms'.[18]

The impetus for this came from De Beers, the monopoly diamond producer that first sought the help of the advertising industry in 1938, as it marked its fiftieth anniversary. It had come far from its origins in the South African diamond rush, set off by the chance discovery of a series of stones, most notably the Star of South Africa, picked up by a shepherd boy on the banks of the Orange River in 1869. The 83.5-carat rough stone, which he traded for 500 sheep, ten oxen and a horse, would eventually come into the ownership of the Earl of Dudley, who paid £25,000 (around £2 million today). More meaningfully, it would spark a frenzy of diamond prospecting in the area – catalyzed by the contemporaneous discovery that the bluish clay characteristic of

---

* Maximilian's marriage to Mary of Burgundy ultimately paved the way for him to become Holy Roman Emperor in 1508. The engagement ring he presented – possibly a surviving jewel designed as the letter 'M' spelled out in diamonds – may have been meaningful, but it did not prove fortunate: Mary died following a hunting accident just five years after their wedding.

what we now call kimberlite pipes could contain rich primary deposits. An initially disparate industry, with limits on the number of mining permits an individual could own, was soon consolidated by a handful of prospectors.

Foremost among them was Cecil Rhodes, who would become a hate figure over a century after his death, with renewed attention on his actions as Prime Minister of the Cape Colony in the 1890s, including his promotion of racial segregation and the murder and war crimes committed under his authority during the conquest of modern Zimbabwe. In the 1870s, still a teenager, he had started his career in an esoteric manner: buying an ice-making machine and selling the product to the working miners.[19] He subsequently turned to renting out the machinery used to pump water out of mineshafts. But his real goal was to acquire control, relentlessly buying up mining smallholdings. By 1888 he had forced a key competitor into a merger and De Beers Consolidated Mines Ltd was formed, borrowing its name from two brothers whose diamond-producing farmland he had bought a decade earlier. The company had an effective monopoly on all diamonds then being mined in South Africa, a stranglehold it would maintain for over a century.

By the 1930s De Beers controlled approximately 80 per cent of the world's rough-diamond trade, but was nevertheless facing problems. Falling demand during the Great Depression forced it to cut production by up to 90 per cent to prevent prices from crashing.* An attempt to popularize diamond jewellery through a partnership with Chanel had also failed. It was at this point that

---

* The Cullinan Mine was closed between 1932 and 1945 due to the Depression and the lack of a market in luxury goods.

it turned to the advertising firm N.W. Ayer, and the history of the King of Gems took a dramatic turn.[20]

Ayer's work for De Beers began in 1938 and at first focused on boosting the diamond's cultural cachet, purchasing works by artists including Picasso and Dalí to feature in its advertisements. During the Second World War, when industrial diamonds were required for the manufacturing of navigation systems in some submarines and aircraft, the focus shifted. 'My most precious possession ... but grind it to powder if it will help give him clearer vision!' one advert's headline declared.*[21] This work had produced early success – De Beers saw its retail sales in the US increase by 55 per cent in 1941 – but it was not until 1947 that it really struck gold. That was when one of Ayer's female copy-writers, Frances Gerety, devised the timeless tagline, 'A Diamond Is Forever' – words that the author was not initially convinced would be a winner.[22] The slogan was not entirely revolutionary, resting on a far older idea: the connection between the diamond's innate hardness and the concept of something unbreakable, ultimately stemming from its etymology – ancient Greek *adamas*, meaning unconquerable or invincible.

The emotional connection now permanently attached to diamond was an enormous achievement, and Ayer pressed on, merrily inventing tradition with remarkable success. Some ideas about the diamond that now feel deeply ingrained were the product of its advertising: how much was expected to be spent on a

---

* The message was not actually for brides to give up their rings. Gem-quality and industrial diamonds are of different grades, although they emerge from common sources. The advertisement sought to assure women that they were not detracting from the war effort by asking for a diamond, and could even be helping by subsidizing the cost of extracting stones for industrial use.

ring ('Isn't two months' salary a small price to pay for something that lasts forever?'), and using the '4 Cs' of colour, clarity, cut as well as carat weight to assess a stone's calibre. Just as 'A Diamond Is Forever' ensured that every woman would eventually want a product sitting stockpiled in the inventory of De Beers, the commercial education of the '4 Cs' would encourage even the small stones to be sold, as there were other qualities than just size to be desired.*

Glossy magazine pages were just the beginning of De Beers' charm offensive: on its behalf, Ayer doled out diamonds to society figures and film stars so they could be photographed wearing them at marquee events. In April 1947 the firm marked a royal visit to Kimberley from King George VI by presenting a diamond each to Princess Margaret and Princess Elizabeth, the latter by then engaged to Philip Mountbatten (an Ayer memo to De Beers had noted that 'the royal couple could be of tremendous assistance . . . by wearing diamonds rather than other jewels'). As De Beers sought to spread the diamond gospel, film stars and royalty alike were treated as living billboards for the promise of prestige, glamour and romance.

The real genius of De Beers' marketing was how it navigated a shifting diamond landscape, giving the consistent impression that an often plentiful material was both rare and desirable, while educating consumers to admire it in different forms. When the Soviet Union discovered diamond mines in Siberia in the 1950s and De Beers became the sole distributor, it was faced with a glut of smaller stones, most less than half a carat before being cut.

---

* The '4 Cs' were originally coined in the 1940s by Robert Shipley, founder of the Gemological Institute of America, to help his gemmology students assess diamond quality.

De Beers came up with the genius solution of the 'eternity ring': a jewel composed of a line of tiny diamonds that was marketed at married men to buy for their wives as an expression of enduring love.[23] Other remarkable successes included Japan, where a De Beers campaign that promoted the diamond as a symbol of 'modern Western values' caused a sizeable shift: the share of engaged women in Japan given a diamond ring rose from just 5 per cent in 1967 to 60 per cent in 1981.[24] That echoed the astonishing success of the American campaign, which had seen De Beers' wholesale sales rise from $23 million in 1939 to $2.1 billion in 1979.[25]

Their next daring decision was a thoroughly modern one, which this time saw the company taking on a market that risked destabilizing the entire global diamond business. The first big breakthroughs in synthetic diamond production were already being made in the 1950s by General Electric, using the high-pressure, high-temperature technique which effectively squashed down carbon, Superman-style. It was a complicated and costly area of research, with initially very limited production. Improvements to this process and the development of a second system which grew diamonds through the deposition of hydrocarbon gases on to natural diamond 'seeds' (which worked like a blueprint the new crystal could follow) saw production advance by leaps and bounds. By the beginning of the twenty-first century, synthetic diamonds – often pitched as ethically and environmentally friendly alternatives[26] – had become a very real threat to the traditional market of natural diamond producers like De Beers.

In a surprising move, in 2018 the diamond giant announced that it would be launching a new range of lab-grown diamonds – 'not to celebrate life's greatest moments, but for fun and fashion' – at a 90 per cent discount on the cost of the natural product. It was a clear message: in an echo of its earlier successful strategies, once more the company planned to control the market (this time through undercutting cost) and to drive consumer demand.

Synthetics were also seen as a viable alternative to a luxury product with a chequered past and a much darker history. The 2006 film *Blood Diamond* highlighted the realities of such conflicts as the civil war in Sierra Leone (1991–2002) where the illegal trade in conflict diamonds to fund war represented up to 15 per cent of the global market at its peak.[27] Although concerted and international efforts have been made to tackle the illicit diamond trade, including the Kimberley Process brought in for certifying rough diamonds in 2003, the challenge lies in a trade where the main commodity is by its nature perfect for smuggling, untraceable and recuttable, and considered worth waging wars over. Lab-grown alternatives are being sold as one solution.

There can be little doubt that diamonds have been the most heavily and successfully marketed gemstone through history; perhaps no luxury item has had a modern commercial story quite like the diamond. All of its innate aesthetic advantages have been extensively polished not just on the jeweller's spinning wheel, but at the advertising industry's boardroom tables. Yet all the tales professionally spun about diamonds are nothing compared to the real stories of how these gems have moved throughout history, been pursued, expropriated and fought over in ways that would never find their way into a magazine advertisement. Diamonds may have been subject to the smoothest storytelling of any gemstone, but the stone's real history is so engaging that little embellishment is required.

'I HAD NOW "caught my hare", declared Lord Dalhousie, British Governor General of India. He was referring to the Koh-i-Noor diamond, a 186-carat jewel that has been one of the most sought-after and fought-over gemstones in history. Its surrender formed part of the Treaty of Lahore signed in March 1849 by the ten-year-old Maharajah of Punjab, Duleep Singh, as the East India Company completed its conquest and annexation of the formerly independent Sikh kingdom. For Dalhousie, the diamond was the ultimate sign of this victory, a gem he wanted on behalf of Queen Victoria, and whose handover was specifically written into the surrender agreement. 'The Koh-i-Noor has become in the lapse of ages a sort of historical emblem of conquest in India,' he reflected.[28]

Its seizure by the British was the latest twist in a long journey of a gem that had already become synonymous with power and plunder, and whose eye-popping history is exceeded by its even more extensive mythology. According to the authors William Dalrymple and Anita Anand, the first definitive account of the stone comes in the eighteenth century, when it adorned the Peacock Throne, a magnificent, gem-encrusted ceremonial seat commissioned by the Mughal emperor Shah Jahan in 1628. Of the throne's crowning canopy, the Persian historian Muhammad Kazim Marvi wrote: 'On top . . . was placed a peacock made of emeralds and rubies; on to its head was attached a diamond the size of a hen's egg, known as the Koh-i-Noor – the Mountain of Light, whose price no one but God Himself could know!'[29]

Marvi was recording the 1739 invasion of Delhi by the Persian ruler Nader Shah, who among his various plunder carried the entire throne, and its crowning diamond, with him back to Khorasan (modern Iran). Nader Shah, who would be assassinated in his tent eight years later, was just one of many owners whose association with the Koh-i-Noor ended in disaster. Those who have possessed it, Dalrymple writes, 'have variously been blinded, slow-poisoned, tortured to death, burned in oil, threatened with drowning, crowned with molten lead, assassinated by their own family and bodyguards, or have lost their kingdoms and died in penury'.[30] After Nader Shah was killed by his own bodyguard, the diamond was taken by his most senior general, Ahmad Khan Abdali, who forged an empire that is the basis of modern Afghanistan while wearing the jewel on an armband.

That achievement began to crumble under the rule of his grandson, Shah Zaman, who was captured and blinded, but not before he had slipped the precious diamond into a crack in the wall of his cell. In 1813 it returned to India in the ownership of Ranjit Singh, de facto founder and first ruler of the Sikh Empire that would dominate the Punjab for half a century until its demise following the second Anglo-Sikh War, a decade after his death. Often wearing it on an armband, Ranjit Singh treated the Koh-i-Noor as a symbol of his rule and conquest, much as Dalhousie would when forcing his son, Duleep, to hand it over in 1849.

Despite the Koh-i-Noor's undoubted star power, its arrival in Britain did not lead to the acclaim that might have been expected. Its public debut came at the 1851 Great Exhibition, baby and brainchild of Queen Victoria's consort, Prince Albert. The diamond was extensively trailed as the event's symbolic showstopper: 'The blazing arch of lucid glass with the hot sun flaming on its polished ribs and sides shone like the Koh-i-Noor itself,' wrote a reporter for *The Times* of the Crystal Palace, the purpose-built hall that housed the exhibition, on the day it opened.[31] Yet shine is exactly what the famous diamond failed to do. Housed in a protective case that resembled a gilded bird's cage, which kept onlookers some distance away, the Mountain of Light was variously decried as 'not bigger than half a fair-sized walnut' and 'nothing more than an egg-shaped lump of glass'.[32] Several re-designs of the exhibit, to prevent it from being swamped in natural light and instead illuminate it with gas lamps, could not overcome the sense of disappointment that surrounded the intended centre-piece of an event that six million visitors showed up to see.

In this, the Koh-i-Noor was underlining an inescapable truth about the diamond: whatever the size of the stone, its provenance and history, there is nothing more important to the gem's appeal than the way it has been cut and polished. This is a complex, multi-faceted and highly expert process, one that begins not with lasers or cutting wheels, but under the magnifying glass of the 'marker'. Looking into the rough diamond, often through a win-dow that has been polished onto the surface, they must make the critical decision about how the stone can be cut to maximize its value: whether into one large stone, two of equal size, or several of different sizes. Like a sculptor, the marker has to be able to see the final product hiding within the raw material, as well as holding the various valuation trade-offs in their head. And like a plastic surgeon, they then ink the lines of their intended cut onto the sur-face, indicating how the before will be turned into the after.

What follows applies modern tools and technology to a trade that has been practised since at least the fourteenth century – sawing, cleaving, bruting (rubbing one diamond against another to abrade it) and polishing. Computer modelling, laser-cutting and modern machinery add precision to work that began in the Middle Ages with crank-operated spinning wheels known as scaifes, coated in oil and diamond dust, but the fundamentals remain: you still need a diamond to shape a diamond, and it is still a process that can, as with the Cullinan, occupy many months.

As the technology for cutting and polishing has evolved, so

has our appreciation of how best to fashion a diamond to optimize its appearance. Understanding this progression in diamond cutting is a vital valuation tool. Countless times I have analyzed a diamond's cut as part of age-dating a larger jewel. Before the advent of diamond-cutting proper, the point- and table-cuts were the only forms for early diamonds: the former retaining the natural octahedral shape, the latter slicing off the top point to create a basic table facet, and allowing light to enter the stone. Both were notable for their similarity to the diamond's natural crystal shape, and make up the majority of all early Roman and medieval mounted diamonds. As cutting took off and the appetite to unleash the diamond's full potential increased, new cuts were experimented with that added more facets and increased the light coming from the stone. The rose cut, with numerous triangular facets on the top and a flat bottom, became popular in the sixteenth century; the briolette was a doubled-up rose-cut drop, with triangular facets all around.

But then a huge change occurred, shifting focus from lustre to brilliance, from light reflecting not just from the exterior but also from the interior of the stone. In the late seventeenth and early eighteenth centuries, the earliest 'brilliant-cuts' were developed, with a wealth of detailed and symmetrical faceting designed to maximize the amount of light reflecting back out of the diamond as well as off it. It marked a shift whereby the aesthetic appeal of the diamond – rather than its cultural or spiritual significance – was becoming the priority for manufacturers, prefiguring its age of mainstream appeal. The brilliant-cut has taken on various shapes and forms since it was popularized. In the nineteenth century, the 'cushion' or 'pillow', with a squared-off outline and larger facets than a modern brilliant, was devised to maximize the diamond's weight as well as appearance under candlelight, and remains today one of the most popular diamond cuts in the world.*

When the Koh-i-Noor came to London, like many Mughal diamonds, it was in a form of rose-cut, with numerous miniature facets around its sides culminating in a four-faceted 'table' at its apex. After its disappointing reception at the Great Exhibition, Albert consulted both scientists and jewellers on how to enhance its appearance. He ended up spending £8,000 (a hundredth of the equivalent modern cost) on having it recut, by a team of British and Dutch jewellers, who reshaped it into an oval brilliant.

The recutting of the Mountain of Light was a major event, one that attracted crowds outside the purpose-built studio and even featured a ceremonial role for the Duke of Wellington, who helped to make the first cut. The work succeeded in enhancing the brilliance of the Koh-i-Noor, but at the cost of 80 of its 186 carats – over 40 per cent of the diamond's weight – despite the initial promise from the jewellers that this could be preserved: a loss at which Prince Albert 'openly expressed his dissatisfaction'.[33] He had now reshaped a gem that his wife would wear as a brooch and that would soon find its way into the Crown Jewels, as the centrepiece of the ceremonial crowns worn by a succession of female consorts up to Queen Elizabeth the Queen Mother in 1937. The Koh-i-Noor made its last state appearance placed on her coffin at her funeral in 2002.

Through its multiple forms and many owners, the Koh-i-Noor has encapsulated why the diamond has been such a symbol

---

* Although it is the most popular cut for diamonds, the brilliant-cut is almost never used for coloured stones, which cannot reflect light in quite the same way, and are often cut more for weight and colour than purely the reflection of light.

of power and politics, as well as a gemstone of unrivalled qualities. It is an illustration of how famous diamonds so often travel complicated paths throughout their history, and the lengths to which owners will go to get hold of them. It is symbolic of their value as treasures and emblems of conquest, and has become central to the debate on how to handle post-colonial collections. Over the years, Indian, Pakistani, Afghani and Iranian claims have been made for the diamond to be returned to them – in some cases, on behalf of countries that no longer exist. Because of this controversy, the Koh-i-Noor was not worn by the Queen Consort (Queen Camilla) at the May 2023 Coronation of King Charles III.

Yet for all the historical challenges in the diamond industry, its fundamental appeal as the hardest and most brilliant material known to humanity remains unaltered, much as it was when diamonds were first being pulled out of Indian riverbeds over 2,000 years ago. Those unique physical properties have given the diamond not just aesthetic supremacy but a cultural life of its own, as a symbol variously of unbreakable wealth, power and love. They have fed the often unruly race both to extract diamonds where new deposits are discovered, and then to fight over ownership of the most magnificent specimens that emerge. These very human impulses that surround the pursuit of diamonds mean that, much as it has become a sophisticated modern industry, there is something gloriously untamed about the diamond that no amount of technical excellence, pricing standardization and advertising muscle can curb. The diamond is one of earth's most miraculous creations, and the promise it holds is as tantalizing now as it has ever been: that another diamondiferous kimberlite will be discovered, something remarkable may be extracted from it, and a crystal that at first glance appears trash will be turned into one of the world's greatest treasures.

9

Coloured Diamond

The King of Diamonds

*'No other gem I know is so rare as a
real blue diamond.'*

EVALYN WALSH MCLEAN

I HAVE held many remarkable gems in my hands, stones with famous owners, twisting histories and a sense of mystery as captivating as the light that shines out of them. But none had ever affected me like this one. The cushion-shaped brilliant nesting in my palm was a marvel of geology and history in its own right: one of the truly awesome examples of the rare breed that is blue diamond. It was about to go up for auction in what was likely to be the biggest public sale of a coloured diamond ever made. But as I looked at the gem, somehow none of that seemed to matter. I suddenly had no thought for the historic importance of the stone or the price it might fetch. I just stared at it. As I did so, I felt something I had never experienced before in the presence of a gemstone. Not the sense of the gem giving me something – a feeling of joy, a frisson of power or a jolt of recognition – but rather a part of me being taken; of my being sucked back into it. I felt pulled in, as if I were falling into a dangerous lake. It was like a piece of my soul had been extracted and absorbed into the greyish-blue pool of the diamond. Some small piece of me gone for ever.

It is not unusual for the most remarkable gems to induce a physical reaction, and especially so with blue diamonds. A few years later I would be shown one of such astonishingly deep, sapphire-like blue that a rash started to creep across my chest and up my neck – something which the dealers who own it still like to tease me about. But the Wittelsbach was different. No other stone had, or has, left me feeling this way – like I had lost something in the process of falling under its spell. That, having seen it and held it in my hand, in some small way I would never be the same again.

Nor would the diamond. It was about to be purchased for a record-smashing $24.3 million and see its fate alter on the polishing wheel. My encounter with this gem had come at a critical juncture in a history already chock-full of intrigue. The 35-carat, postage-stamp-sized stone had emerged from the Golconda region in India, usually thought to have been the same Kollur mine that produced the Koh-i-Noor as well as some of history's most famous coloured diamonds. Another of the most prominent blues (the Hope diamond) and the foremost of the greens (the Dresden Green), as well as a number of historic pinks (including the Darya-i-Noor and Noor-ul-Ain), are all believed to have come from this source.

Records of the Wittelsbach begin in the seventeenth century. Although the familiar narrative that it was purchased by King Philip IV of Spain as the dowry for his daughter, Infanta Margarita Teresa, has been brought into question, it has been established that she likely did own the diamond. She bequeathed it to her husband, Holy Roman Emperor Leopold I, after her death in 1673, aged just twenty-one.[1] By inheritance and inter-marriage, the stone enjoyed a royal progress through the Spanish, Habsburg and ultimately Bavarian dynasties, where it became established as a Crown Jewel – most notably on the Royal Crown of Bavaria after the former electorate became a monarchy in 1806. It was from the House of Wittelsbach that the diamond took its name and in its hands that it enjoyed its longest period of ownership. Not until the twentieth century did it emerge onto the open market: a failed attempt was made to sell it in 1931 to raise funds, and only in 1951 did it finally leave Germany.

Ironically, it was in the hands of the jewellery trade that this almost matchless gemstone threatened to become lost and its famous history obscured. A prominent Belgian diamond dealer had bought the stone but made so little fuss about it – including exhibiting it anonymously at a major exhibition in 1958 – that on his death his family appear not to have recognized what it was. Only when they took it to another Antwerp-based jeweller for recutting was its real identity detected. That jeweller was Jozef Komkommer, and instead of drastically reshaping the diamond as asked, he recognized its heritage and bought it from his would-be clients, safeguarding its shape and significance. Having failed to sell it back to what remained of the House of Wittelsbach, he found a buyer at the end of 1964. This was Helmut Horten, a German retail entrepreneur, who presented the diamond to his new wife, Heidi, at their wedding party in Cap D'Antibes in 1966. It was through her that the stone came to Christie's some forty years later in 2008, and into the palm of my unsuspecting hand.*

Many gemstones have had notable historical journeys, including prominent owners and enticing tales of being lost and found.

---

* The highlights from the rest of her jewellery collection came up for sale in 2023, but not without controversy when it was revealed that her husband's wealth had in part been amassed by purchasing Jewish businesses sold under duress in the 1930s. Despite a boycott from several prominent members of the trade, the auction set a world record for a single-owner jewellery sale of over $200 million.

So what made this diamond so special that it attracted one of the biggest bids ever made, and left me feeling so profoundly affected? Quite simply, it was the colour. Diamonds are remarkable things to look at and so, in different ways, are coloured gemstones. But nothing comes close to the effect achieved when you put the two together – the deep emotional associations of colour, dazzling at maximum brilliance thanks to the close-packed cubic carbon structure of the diamond, and its unique light-bending properties. There is no mystery at all about why the list of the most expensive gemstones ever sold is largely a list of coloured diamonds.* They are the brightest and the best, the rarest and the most remarkable: all the most desirable qualities of a gemstone brought together, every mind-altering trick of light and colour at their disposal.

Coloured diamonds come in almost every hue imaginable, from the commoner yellows and browns, to the routinely record-breaking pinks and blues, and outrageously rare oranges, greens, purples and reds – the latter being so scarce that they are seldom seen on the market. Their value is directly linked to their colour, usually assessed by the global authority in diamond grading, the Gemological Institute of America (GIA). Measured on a scale ranging from light to fancy, intense, vivid, deep and dark, the GIA's colour-grading system encompasses both the tone (relative darkness) and saturation (strength) of the colour.

Blues are not the rarest of this bunch, but they have often been the coloured diamonds to break records and fire imaginations. These are the Type IIb diamonds, given their colour by boron, that

---

* The Wittelsbach was a record-breaking coloured diamond at the time of its sale in 2008. That record has been exceeded numerous times since by over a dozen pink and blue diamonds and in one case an orange – testament to the strong market for these stones in recent years.

represent just 0.1 per cent of the total supply. Stones such as the Wittelsbach, the Hope, the Oppenheimer Blue and the Blue Moon of Josephine are the select few of an already tiny subset. Some are historic and have passed through the hands of many owners, whereas others have been relatively recently pulled out of the earth by modern mining machinery. Regardless, they are the unicorns of their breed, around which enigmatic histories, celebrated owners and eye-watering price tags have clustered. As I experienced with the Wittelsbach, there is something tangibly evocative about boronic blue that adds yet another layer to their allure. While they can reach the sky-high heights of baby blue, most emotive are often the deeper, darker, more sea-like stones. Older gems in particular, cut for weight over colour, often retain a softness to their tone, a watery quality that recalls the depths of the ocean or a dark pond. It was into just such a deep pool that the Wittelsbach had dragged me, making me feel like I had left a part of myself behind.

And now this celebrated stone was about to change – not just ownership but size, shape and even name. The 2008 sale did not just attract a mammoth price tag. It also passed the Wittelsbach diamond into the hands of a new and highly significant owner, the jeweller Laurence Graff, one of the most influential figures in the industry and a legend in the field of coloured diamonds. The decision he made to rework the stone would bitterly divide the trade, revealing the fierce debates that surround these gems as well as reinforcing the central importance of the cut to any diamond's story.

In terms of weight, the impact of this surgery was only slight, slimming down the stone from 35.56 to 31.06 carats. Yet the effect was remarkable. The gentle, soft, greyish tones that my eyes had swum through were gone, replaced by a much purer, cleaner blue. The culet – the bottom facet of the diamond – had been reduced, allowing for more improved light-giving facets around the base. By updating a stone that had probably last been cut in the seventeenth century and applying modern techniques, Graff had given his famous diamond the gift of perfect clarity, stoked the stone's 'fire', and enhanced its ability to throw out reflections. Less light was seeping out of the back of the stone – the silent sucking away at my soul, that had given rise to my feeling of somehow being spiritually robbed. Now it was being brilliantly reflected back for the benefit of the observer, making the diamond about as bright as could be. The Wittelsbach–Graff doesn't just look more lively and tidy than its former incarnation. It also appears distinctly bluer. With the lack of light, so went the grey. On its own terms, this was a masterpiece of the diamantaire's art, showing once again how much of the diamond's appearance depends on the quality of the cut and polish.

The Wittelsbach's evolution into the Wittelsbach–Graff was remarkable, but also hugely controversial. Many dealers, curators and connoisseurs cried foul and declared a scandal. For them, the refashioning of such an iconic diamond was a desecration, mutilating a valuable historic object. One museum director complained that the stone had been 'turned into a piece of hard candy' and said that it was akin to painting over a Rembrandt.[2] A fellow diamond-cutter declared it 'barbarism', saying: 'You cannot begin to describe the damage he has done. The stone may look a bit more lively and sparkling, but its history has been destroyed.'[3] Anger was particularly apparent in Germany, with the feeling that a piece of national history had been lost. The *Frankfurter Allgemeine Zeitung* even published an obituary for the stone, describing its recutting as the 'abolition of eternity'.[4] Other complaints centred around the fact that Graff had also slapped his name on the stone, and that he had carved up a historic piece purely to increase the market value of the object – which the reworking certainly had done.* It was hard cash over history, critics lamented: an ugly symbol of commerce's triumph over art.

As hypnotized as I was by the original stone, I could still see two sides to the story.† Embracing the history of gemstones also means accepting how central the process of change has been to that history. Almost every notable stone has changed ownership, changed use and changed setting on multiple occasions. Many of them have also changed form – being recut into new shapes, or having names engraved onto them. Few diamonds ever retain their original form: the roughs that emerge from the mines are only special because of what human hands do to cut, polish and market them. All brilliant diamonds have been worked. So at what stage should the manufacturing of a stone be declared complete, when time and technological developments only offer us better ways of showcasing these marvels of nature?‡

---

* Trade estimates for the new value of the stone at the time ranged from $50 to $100 million.
† The debate about preservation was one I had first encountered as an archaeology student. On my first dig in Pompeii, I was told to smash up an original Roman *cocciopesto* pavement in order to access the earlier inhabitation layers beneath, and point blank refused. The discussion that followed taught me a great deal about the need to record and sometimes remove history in order to further research, but my stance saw me moved to excavating the gutters and latrines of the house instead. A point well made by my supervisor, I assumed, to prove the transience of human habitation.
‡ Which is not to say that all modernizations are created equal, or to be welcomed. The Agra Diamond, one of the largest historic pink diamonds, with its origins in fifteenth-century India, has rarely been seen since a radical recutting after the last of its many sales in 1990. This work succeeded in enhancing its colour, but at the cost of losing its lovely old cushion-shape in favour of a modern overly faceted cut-cornered rectangular radiant-cut: a far more extreme change in shape than was the case with the Wittelsbach.

In the case of the now Wittelsbach–Graff, the diamond had also been badly bashed about from centuries of wear and had inherited chips and bruises around the edge of its ultra-fine and very irregular girdle – the thin facet running all the way around the edge of the stone. By tidying those up and reducing the culet, the diamond reappeared far less damaged, but entirely recognizable – still clearly the historic Wittelsbach, just neatened up. The integrity and outline of the stone had been retained, while improving its beauty and overall appearance: something that could only have been done by the best in the business – both the most experienced and audacious. Perhaps the knocks and chips that the stone had picked up were signs of a life excitingly lived, but so was this next chapter. Its newest incarnation would now also become a part of its history.

This humanity is a large part of what makes the diamond so captivating as an art object: the combination of natural beauty, worked by human skill, for aesthetic improvement and financial interest. The moment a stone comes out of the earth and into our hands, it has entered a new, living journey, no longer at the whim of its earthly environment, but answerable to those who now own it. More often than not, it becomes part of a shifting industry driven by finance, profit and success. It is human nature to want to make a mark on the world. And when it comes to legacy, reworking and renaming a diamond is about as eternal and enduring a legacy as one can hope for.

In the end, diamonds are dynamic, portable and practical objects, not listed buildings or Old Master paintings.* The truth

---

* Although as Graff responded to his detractors, 'If you discovered a Leonardo da Vinci with a tear in it and covered in mud, you would want to repair it.'

about gems has always been, whether humans are fighting over them on the battlefield or bidding for them in an auction room, that to the victor go the spoils. Like it or not, the owners of a gemstone get to determine its fate. Those who wish to protect a stone from the cutter's wheel can bid for it on the open market just like the competitors who dream of remaking it. This time, the victor was Laurence Graff, and his name would become part of its updated history.

Whether you consider the Wittelsbach–Graff to have been impaired or enhanced by its recutting, the storm over its modern incarnation was trivial in the context of coloured-diamond history. These are not just the most valuable and, in many ways, desirable gems. They are also the standout stones in other ways – the ones most overshadowed by curses, controversies and outlandish stories. The history of the coloured diamond is everything you thought you knew about gemstones, with the volume turned up: more epic, more extraordinary, more expensive. When you are holding one of these prizes in your hand, it is not just the cut and colour that take your breath away, but the story of how it got there. These narratives are the most lavish surrounding any gemstone, just as the diamond's colour and brilliance stand supreme. They are tales of theft and adventure, of battles legal and financial, of great riches and dramatic falls from grace. And the best part? Some of those stories may even be true.

I N EARLY 2010, shortly after the work on the Wittelsbach–Graff had been completed, it went on display at the Smithsonian Museum in Washington DC. This was no ordinary exhibition: it showcased not just one but two of the most famous blue

diamonds ever discovered. Alongside the freshly polished Wittelsbach was a stone that had been in the museum for half a
century, one that had experienced several lifetimes' worth of adventure before ending up in a display case. This was the 45.52-carat
Hope Diamond – perhaps the most storied stone of them all,
and certainly the gem that epitomizes best of all the twisted history of the coloured diamond. When gifted to the museum in
1958, it had arrived after being sent through the post in a brown
paper bag – 'the safest way to mail gems', according to the man
who was giving it away, the legendary diamond dealer Harry
Winston.[5] This final, humble journey was in stark contrast to all
those that had preceded it. No stone has travelled across continents and between owners with more mystery and misfortune
than the Hope. No gem better shows the coloured diamond's
close association with curses and calamity – the idea that, behind
the face, there is more than meets the eye.

Its story begins with one of the most important figures in
coloured-diamond history – the French merchant and explorer
Jean-Baptiste Tavernier. Tavernier travelled widely in India and
Persia during the seventeenth century, trading some of the most
magnificent gems in the world. Among his illustrious patrons
was the Sun King, Louis XIV, who also pressed him to commit
the story of his six voyages to print. Tavernier's journeys through
India, the first commencing in 1631, were genuinely epic. He
travelled an estimated 20,000 miles in total, over a period of
thirty-eight years, suffered shipwreck, and at one point was
thrown in jail.[6] He was the inveterate traveller, constantly out on
a quest to uncover the diamonds he had read about in the *Arabian Nights* and the works of Marco Polo.

Tavernier was rewarded for his intrepid spirit and assiduous
cultivation of local contacts with the chance to see some truly

remarkable diamonds. In 1642 he encountered – though was unable to purchase – the Great Table, a pink diamond whose moniker was no exaggeration, for it measured almost 60 millimetres across. A Mughal stone that was one of those taken by Nader Shah when he sacked Delhi in 1739, its pale pink would later be cut into two stones that remain part of Iran's Crown Jewels: the 60-carat Noor-ul-Ain ('Light of the Eye') and the 182-carat Darya-il-Noor ('Sea of Light'). The Noor-ul-Ain was mounted in the wedding tiara of Empress Farah Pahlavi by Harry Winston, for her marriage to the Shah of Iran in 1958. The rectangular table-cut of the Darya-il-Noor mimics the original stone and is engraved with the name of the long-reigning Shah of Iran, Fath-Ali Shah Qajar (1797–1834). It remains the largest pink diamond in existence.

It was on Tavernier's final voyage to India, beginning in 1663, that he encountered something miraculous that he was able to buy: a 115.16-carat blue diamond – *'net et d'un beau violet'* (clean and a beautiful violet – i.e. flawless and deep blue).[7] Most likely it came from Kollur, the Golconda region's proven producer of showstopping coloured diamonds. Frustratingly, despite the extensive chronicle he recorded of his adventures in India, which included a drawing of this stone, there is no account of how it came into his possession. Tavernier may have acquired it from one of his contacts at the mine, or perhaps from a merchant elsewhere on his travels (there is a record of him purchasing an extremely expensive diamond on that trip while passing through Isfahan, in modern Iran).[8] This ambiguity over its provenance would lead to subsequent mythmaking at Tavernier's expense, suggesting it was acquired in more illicit circumstances than are credible. All we know for sure is that, in December 1668, he met with Louis XIV and his powerful finance minister, Jean-Baptiste

Colbert, and sold a job lot of over one thousand diamonds including his star find – described by court records as 'a large blue diamond in the shape of a heart, thick, cut in the Indian style'.[9] The Sun King quickly set about making his new prize shine, ordering his court jeweller Jean Pitau to recut it. Since it had initially been faceted to the contemporary Indian standard, 'to preserve weight at the expense of symmetry and brilliance', this was a fairly drastic job. The stone was slimmed down to approximately sixty-nine carats, into a shield-shaped brilliant. Colbert named it the *Diamant Bleu de la Couronne* (Blue Diamond of the Crown), and the stone would subsequently become known simply as the French Blue.[10] It would remain in this form until the arrival of the French Revolution brought with it the beginning of a new and uncertain chapter.

That began in September 1792, when the *Garde-Meuble* (Royal Storehouse), whose contents had been turned over to the Revolutionary government the previous year, was ransacked. During what turned out to be nearly a week of nightly looting, one of the greatest robberies in the world unfolded. The French Crown Jewels, confiscated from the Crown during the chaos of the French Revolution, were gone. While the majority were recovered in the weeks and months that followed, the French Blue was not, and it would never be seen in the same form again. In fact, it does not reappear in any reliable record until 1812, when it was documented in the possession of a London diamond merchant, Daniel Eliason, having been cut down again to its present weight (allowing for a little later repolishing).[11]

It was probably Eliason who sold it to the diamond's next confirmed owner, and the one to whom the gem owes its current name, the banker Henry Philip Hope. This financier may have been renowned as 'munificent in his charities',[12] but he was also a

The Imperial State Crown, mounted with the 317-carat Cullinan II diamond, the 170-carat Black Prince's Ruby (a spinel), and various other diamonds, rubies, sapphires, emeralds and pearls, and made for the coronation of King George VI in 1937, credit: Royal Collection Trust/© His Majesty King Charles III 2023.

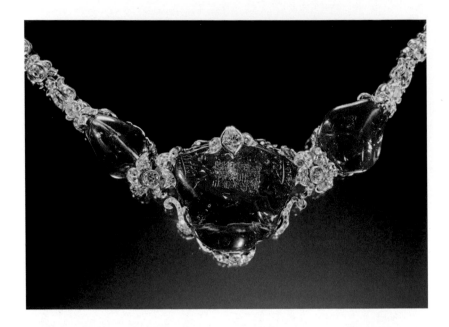

The Timur Ruby Necklace, set with a 352.5-carat spinel, two further spinels and diamonds, and mounted by Garrards in 1853, credit: Royal Collection Trust/All Rights Reserved; a Mughal spinel necklace of eleven polished spinels, 1,131.59 carats, three engraved © 2011 Christie's Images Ltd; the Carew Spinel, 133.5 carats, also an engraved and polished bead © Victoria and Albert Museum, London; the step-cut Hope Spinel, 50.13 carats, set in a nineteenth-century brooch, credit: Bonhams.

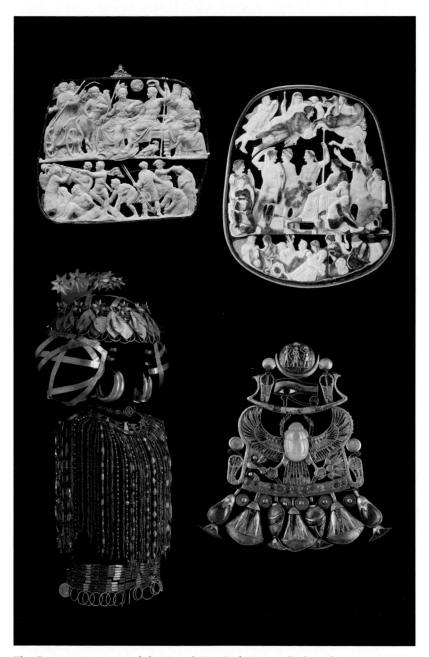

The Gemma Augustea and the Grand Camée de France, both sardonyx cameos, first century AD, © KHM-Museumsverband, ©BnF; Queen Puabi's gold headdress and beaded carnelian, agate and lapis lazuli cape, *c.* 2500 BC, courtesy of Penn Museum © 2015 Bruce White; Tutankhamun's pectoral ornament, set with carnelian, lapis lazuli and turquoise, and desert glass, 1323 BC, credit: De Agostini/A. Jemolo/agefotostock.

The nine major stones cut from the 3,106-carat Cullinan Diamond, including the Cullinan I (top centre: 530 carats) and the Cullinan II (top left: 317 carats); the diamond-set Poltimore Tiara, made in 1870 for Lady Poltimore and purchased at auction in 1959 for Princess Margaret, credit: *Tiaras: A History of Splendour* by Geoffrey C. Munn (ACC Art Books).

Some of the vivid hues of fancy coloured diamonds © Larry J. West; the 41-carat Dresden Green Diamond, mounted in a diamond hat ornament of 1768, courtesy of SKD, photo: Carlo Böttger; the 31.06-carat deep blue Wittelsbach–Graff Diamond, courtesy of Graff; the 45.52-carat deep greyish-blue Hope Diamond, courtesy of the Smithsonian, photo: Chip Clark; a selection of pink and red diamonds from the Argyle Mine in Australia © Rio Tinto.

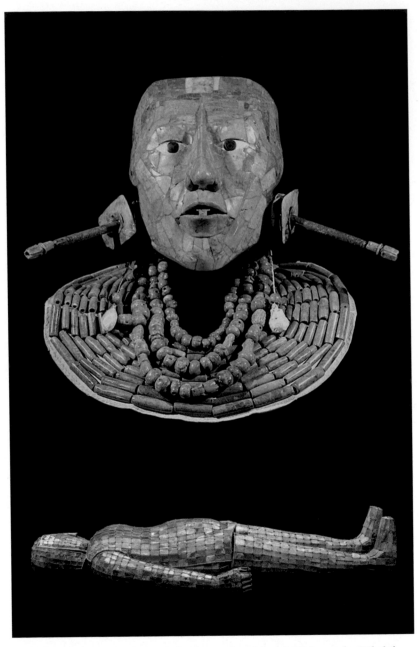

The jade death mask (with ear perforators and necklaces) of Maya ruler Pakal the Great, *c.* 683, Palenque, National Museum of Anthropology, Mexico City; the jade burial suit (stitched together with gold thread) of Liu Sheng, Prince of Zhongshan of the Western Han empire, 113 BC, photo by Zhang Peng/LightRocket via Getty Images.

'Mutton fat' white jade wine cup of Shah Jahan, 1657 © Victoria and Albert Museum, London; portrait of Pare Watene holding a jade mere club and wearing a jade hei-tiki necklace, by Gottfried Lindauer, 1878, credit: Auckland Art Gallery, gift of Mr H. E. Partridge, 1915; Barbara Hutton's jade bead necklace, on a ruby and diamond clasp, Cartier Paris, 1934, Marian Gérard, Collection Cartier © Cartier.

The Duchess of Windsor's blue chalcedony, sapphire and diamond suite, including a two-strand necklace with flowerhead clasp, and a pair of couronne bangles, made by Suzanne Belperron, *c.* 1935, courtesy of Siegelson.

seasoned gem connoisseur who knew how to drive a hard bargain. One account of the sale has him writing out a cheque and putting it on the table along with his watch, telling Eliason that the offer would only stand for five minutes. The dealer accepted, complaining that he was getting the prize 'dog cheap'.[13] Hope perhaps knew that Eliason had been hawking the gem around Europe without success, and reportedly closed the deal for little more than a third of the original asking price. If the purchase did indeed take place on these terms, it portended a future for the Hope in which it would repeatedly be haggled over and sold in contentious or distressed financial circumstances. Hope's legatees, his three nephews, fought a lengthy court battle in the 1840s to establish which of them would inherit the diamond, while the family's final owner – Lord Francis Hope – flogged it in 1901 to help service his heavy debts.

Its next notable guardian, nearly a decade later, was the house of Cartier. Having opened its first New York store months earlier, Pierre Cartier sought out the Hope as a signature purchase to help put his firm on the map with high-rolling American clients. One already known to him was the heiress and socialite Evalyn McLean, who had both inherited and married into wealth. She was the daughter of Thomas Walsh, an Irish immigrant who became one of the era's most successful gold miners, prompting the title of her memoir, *Father Struck It Rich*. Evalyn cemented the family's new-found wealth and social status by marrying Ned McLean, heir to a publishing empire that included the *Washington Post*. With two fortunes at their disposal, the McLeans lived and spent seemingly without limits. They had already bought one outrageously expensive diamond from Cartier – the 94.75-carat Star of the East – while honeymooning in Paris in 1908 ('I need no guide to find that place,' Evalyn later wrote of the jeweller's flagship boutique on Rue de la Paix).[14] Now they were being lined up as potential buyers for the Hope.

Pierre Cartier knew his customers, and he courted them assiduously, arranging to meet but without disclosing the treasure he had to offer them. 'His manner was exquisitely mysterious,' McLean recalled in her memoir. 'I suppose a Parisian jewel merchant who seeks to trade among the ultra-rich has to be more or less a stage manager and an actor. Certainly he must be one great salesman.'[15] Cartier did not disappoint: without opening his wax-sealed parcel to reveal the treasure contained within, he launched into an extensive and often fictitious account of the stone that had been the French Blue and become the Hope Diamond.*[16] He recounted how Tavernier had 'stolen the gem from a Hindu, perhaps a Hindu God' and subsequently met an unfortunate end, torn apart by wild dogs; how Marie Antoinette had worn it before her untimely date with the guillotine; how the Hope family fortune had dwindled under its spell; and how a subsequent owner – the Turkish diamond dealer Selim Habib – had drowned at sea after selling the gem. This intricate tapestry of truth and legend achieved its desired effect: a dealer who had come with something to sell now had a customer champing at the bit to view the goods. 'Let me see the thing,' McLean finally exclaimed to Cartier, who apparently waited a full minute – 'as a concert pianist may do before striking trained fingers to the keys of his instrument' – before finally revealing it.[17]

Despite having enticed his client with this intricate sales

---

* It was always assumed, but only confirmed by computer modelling in 2005, that the Hope had descended from the French Blue. A subsequent study debunked theories that the Wittelsbach was an offcut from the same origin.

pitch, she turned him down. 'Ned – I don't want the thing. I don't like the setting', she declared. Cartier was not so easily deterred. He had the Hope remounted onto a diamond necklace and pressed her to spend a few days with it, correctly betting that familiarity would breed content. The gambit was successful, and the McLeans agreed to purchase the diamond – a transaction that would be far from straightforward. A price of $180,000 (roughly $5.5 million today) was agreed and a contract signed, but weeks passed with no initial payment being made. Then Evalyn tried to back out of the deal altogether, under pressure from her mother-in-law, who insisted she return the 'cursed stone'.[18] Cartier eventually had to resort to lawyers to ensure his clients complied with the contract they had signed. It was over a year before the matter was settled, and the delays combined with legal fees left the firm out of pocket. Cartier had been caught out by his own 'curse'.

For Evalyn Walsh McLean, her ownership of the diamond only seemed to confirm its connection to misfortune. Although she claimed that 'I like to pretend the thing brings good luck', in the years that followed her marriage collapsed and her husband died in a psychiatric facility, one son was fatally hit by a car, her daughter died of a drug overdose, Evalyn herself was defrauded by a con artist who fleeced her of $100,000, and her finances became so strained at one point that she had to pawn the Hope. The diamond was reclaimed and remained in her ownership until her death in 1947, but it was then sold along with the rest of her jewellery to New York jeweller Harry Winston to help cover her estate's heavy debts.

Evalyn had bought the Hope Diamond partly because of its twisted history, and she appeared to have become a poster child for the idea that it was in fact cursed. Although many of the

Hope's owners seemed to meet with misfortune in one way or another,* some of the best stories in the stone's supposedly ill-starred history fall apart on cursory inspection. The founding legend – that Tavernier brought about the curse by stealing the diamond from an idol in a Hindu temple, and was subsequently savaged by wolves – was preposterous: he was a highly respectable trader of long standing who lived to the ripe old age of eighty-four.[19] Nor was the line that Cartier spun about Marie Antoinette having worn the stone at all likely, since it was mounted in the ceremonial Order of the Golden Fleece at the time.[20] The tale of the Turkish diamond dealer drowned at sea was a case of mistaken identity: it was a different man with the same name who went down on the ill-fated paddle steamer. The story of the curse is one that liberally mixes fabrication in with fact, and lumps together those who actually possessed the stone with those who almost certainly never touched it.

Still the curse story persisted – then and now – because there is some part of us that wishes to believe it, or at least to entertain it. When Pierre Cartier had tantalized Evalyn Walsh McLean with a heady cocktail of sparkling history, seductive anecdotes and entertaining speculation about the Hope, she both recognized the game being played and could not help herself from participating in it. 'Do I believe in a lot of silly superstitions, legends of the diamond? I must confess I know better and yet, knowing better, I believe.'[21] She ensured that her children never touched the Hope and at one point even took it to a Catholic

---

* Even the postman who brought the diamond to the Smithsonian, James Todd, seems not to have escaped: shortly after making his famous delivery a truck ran over and badly injured his leg, and later his house burned down.

priest to be blessed (appropriately enough, a storm descended as he did so, lightning splitting a tree in half).[22]

Her behaviour is eloquent about the effect of such stories on our psychology, and how they can become such an important part of the desirability of gemstones. Aesthetic quality and associations with famous owners past may be attraction enough. But there is also a magnetism in mystery: the question of where a stone really came from, in whose hands it has actually been held, and if it could truly carry a power beyond beauty. Like the tendrils of a weed, these thoughts are hard to banish once planted. They are pondered, even without being fully believed. And over time, as coincidence piles up on happenstance, they add a layer of sheen to a gemstone that even months of dogged polishing cannot.

WHEN EVALYN Walsh McLean was examining the Hope Diamond for the first time, she found herself both enticed and wrongfooted by its colour. 'The blue of it is something I am puzzled to name,' she later wrote. 'Peking blue would be too dark, West Point blue too gray. A Hussar's coat? Delft? A harbor blue?'[23] She is not the only one to have looked at a coloured diamond and to have felt that they were experiencing something novel and almost indescribable. The hue of these diamonds is central to their aura – entrancing not just because they are beautiful, but also unfamiliar and somehow intangible. So it is appropriate that these colours should have come variously from more profound and mysterious origins than other gemstones. The story of how diamonds got their colour is one that takes us deeper into the earth's layers than any other gemstone, and whose details geologists continue to puzzle over.

In the case of blue diamonds, we know that the colour comes from boron, but the difficulty arises from the fact that this element – which is abundant at the surface of the earth – barely exists at the depth where diamonds form. Because gemmologists rely on inclusions to deduce the origin of gemstones, and in coloured diamonds these have often been eliminated by the cutter where they even existed in the first place, for a long time it proved impossible to establish where exactly these enigmatic gems had emerged. Only in 2018 did a study of forty-six blue diamonds with inclusions shed some light on this mystery.[24] The minerals they contained helped researchers to deduce that these diamonds had formed even further down than previously believed – as deep as the earth's lower mantle, 660 kilometres below the surface, compared to the 150–200 kilometre depth at which diamonds are typically known to form. The boron that gave them their colour is thought to have been carried down all that way by pieces of the ocean floor that had undergone subduction. The high heat and pressure of the lower mantle then melted this rock, releasing the element that colours the diamond blue. These remarkably rare gems, therefore, were brought to us by one of the earth's most magnificent rollercoaster rides: the great descent of whole chunks of ocean, a precursor to the ascent that brought the diamonds exploding to the surface.

When it comes to wild and wonderful origin stories, blue diamonds must share the limelight. The pinks have a curious story of their own, their tone arising not from any mineral intervention but from a quirk of structure. Here heat and pressure do all the work, stretching the bonds between a diamond's carbon atoms in a process known as plastic deformation. The diamond's structure is effectively squashed out of shape, not unlike the effect of squeezing a pack of playing cards until they each slide out from under one another: a physical defect capable of causing a beautiful anomaly. Depending on the type and origin of the diamond, this process can create colours ranging from the pale pink of the Agra or Darya-i-Noor, to the incredibly rare red, and others that cover a spectrum of orange and brown tints.

Greens, too, share a weird physical fault. Their colour is caused by radiation, which knocks the carbon atoms out of place in the crystal structure, also altering the light absorption. While this can occur within the earth, artificially irradiated stones are far more common. The natural process is so difficult to distinguish from man-made irradiation treatments that, unless the levels are off the scale, it can at times be almost impossible to distinguish a genuine green from an irradiated stone.* The most valuable greens, such as the Dresden Green, are those known to pre-date the invention of laboratory-induced irradiation.

What unites these various processes is how unusual they are. While diamonds are not a rare gemstone, those containing colour make up a minuscule fraction of the total. The GIA estimates that they represent just one in every 10,000 carats of fashioned diamonds in existence. Those with what it calls an intense colour are even rarer: one in 25,000.[25] For most of their history, this meant that coloured diamonds were vanishingly rare finds, secreted into the safes and treasuries of the most exclusive collectors. There was no mainstream market in them because they simply did not exist in any meaningful number, appearing

---

* I have dealt with a 100-year-old stone that – although buried in the owner's garden for so long that it had to be natural – could not be authenticated by the laboratory. And I once sent off a green diamond for testing only to have it returned encased in lead because so much dangerous artificial irradiation had been detected.

anomalously in diamond mines ranging from India and South Africa to Brazil, Russia and Tanzania.

One unusual deposit is the exception that proves the rule. In 1979 a diamond pipe was discovered in Western Australia which would become the Argyle Mine. For the first time, a reliable source of coloured diamonds had been unearthed: according to one report from 2001, just 27 per cent of the diamonds emerging from this mine were colourless or pale yellow, whereas 72 per cent were brown and the remaining 1 per cent a mixture of blue, green and predominantly pink stones.[26] For pinks in particular, this marked a revolution. Even as such a small proportion of the total deposit, they were now being extracted with unprecedented regularity. While still extremely rare, they started to become, for the first time in history, available. Argyle pinks (and even some reds) created a high-end market of their own, a competitive arena for a handful of committed collectors with the means to match their interest.

Although Argyle was not producing large enough gems to enter the international auction headlines, its existence certainly supported the rise of pinks on their upwards trajectory. Between 2012 and 2022 five of the ten most expensive gemstones ever sold under the hammer have been pink diamonds, at prices ranging from $39.3 million to $71.2 million. The price of pinks experienced a steady surge as stones from the Argyle Mine started to come onto the market, then as those that had been sitting in safes for years emerged to catch the rising tide, and finally as it became known that the Argyle Mine was preparing to shut down, which it finally did in 2020. Jaw-dropping amounts were expended: in 2015 the billionaire Joseph Lau parted with $28.5 million to buy a vivid pink diamond at Christie's for his seven-year-old daughter, naming it the Sweet Josephine. At Sotheby's, the very next day, he bid $48.4 million for a 12.03-carat vivid blue, which became the Blue Moon of Josephine, also taking her name.[27]

Pinks did not pioneer this coloured-diamond extravaganza alone. While Argyle's brown diamonds were considerably more abundant and less desirable, the mine's promoters nevertheless applied artful marketing to add lustre to a previously unloved variety. Brown became 'champagne', 'cognac' and 'chocolate', and the category later benefited from celebrity endorsers including the actress Scarlett Johansson, who chose one for her engagement ring.

More valuable are the yellow diamonds whose nitrogen content gives them a sufficiently strong colour to be distinguished from those regarded as a subset of the colourless diamond. Although these are the most common of coloured diamonds, there are some famous and stunning examples. Foremost among them is the Tiffany Diamond, a 128.54-carat yellow that emerged from South Africa in 1877, distinguished by the fact that just four women have ever worn it – Mary Whitehouse (the wife of an American diplomat), Audrey Hepburn, Beyoncé and Lady Gaga. Coloured-diamond king Laurence Graff has also been a consistent promoter of yellows. Since buying the 47.39-carat Star of Bombay in 1974, a historic stone from the Golconda mines, he has brought a long succession of these diamonds to market and has helped to drive up their popularity and price: in 2014 he achieved a $16.3-million price for the Graff Vivid Yellow, a 100.09-carat diamond that had come from a rough almost twice as large.[28]

The rise and rise of the coloured diamond owes much to the sophistication of the modern jewellery industry. Advances in mining have seen larger specimens recovered in recent years. Expertise in cutting means that these unmatchable stones, with their peerless interplay of colour and brilliance, can now be seen in their best possible light. Although some of these facelifts have been controversial, they also prove the near alchemy involved in cutting a stone to change not only its shape but its actual colour. In turn, the astute marketing that has been part of the diamond's story since the mid-twentieth century has been hard at work in uplifting the less obviously attractive of the fancy varieties.

These market forces have driven the popularity of a stone that is perhaps the most remarkable natural creation of all – the apex gemstone that combines richness of colour with the diamond's enrapturing reaction to light. The coloured diamond is an extraordinarily rare and truly remarkable object. But it also speaks to universal truths about gemstones – commodity objects whose worth can never be wholly financial or entirely quantifiable by normal measures. While flows of capital help to define the direction of prices, aesthetics, history and emotional resonance are the factors that ultimately enshrine value.

At the start of my career in the auction business, I learned to value gems with my head, using price per carat benchmarks. The more I watched experienced dealers, the more I realized that the best make equal use of gut and heart. While you must know what the material is worth functionally, you also need to be able to intuit what it will mean to people. You have to be confident that, if you have fallen in love with a stone, others will too. Even allowing for differences in individual taste and cultural preference, there is a surprising degree of overlap in what attracts people to these objects. The emotions spurred by the beauty of a gorgeous gem, the associations ingrained in its colour and the attraction created by its story are so often universal. And if you want to understand how these factors can drive simple lumps of carbon to unimaginable heights of value, then you need look no further than the inscrutable, unbelievable and often near unattainable marvel that is the coloured diamond. It is the gemstone that both stands alone in quality and tells the human story of why these objects are so widely loved and pursued – as symbols of love, slices of history and sources of a fascination that feel as eternal as the diamond itself.

10

Jade

The Stone of Heaven

'A noble person's virtue is like jade.'

CONFUCIUS

I F THERE WAS one gem material that I was never really expected to appreciate, let alone understand, that would be jade. Not a material prominent in our Western canon, it is missing from the ancient and medieval European lapidary texts, and only pops up in high jewellery at rather scattered and specific moments in modern history. For many of us in the West, jade is simply not on our radar.

Where it is understood, however, jade rules supreme. To the Chinese, it is part of their culture and in their blood. They have a saying: 'Gold has a value, jade is invaluable.' It expresses a notion that is intuitive in some parts of the world, yet absolutely alien to others.

When the last Aztec Emperor, Montezuma II, was negotiating with the Spanish invader Hernán Cortés in 1519, Montezuma conveyed a similar message. He presented Cortés with several jades, in his own words, 'of such enormous value that I would not consent to give them to anyone save to such a powerful emperor

[Charles I of Spain] as yours: each one of these stones is worth two loads of gold.'[1] The conquistadors initially wondered if they were emeralds. But they soon learned that these green and white pebbles were worth far more as they were the perfect bartering tool to build alliances – with other tribes tiring of Aztec rule – that helped pave the way to their eventual military victory. This gift symbolized how the value of jade outside its homelands has had to be learned and assimilated by outsiders.

In 1870 in New Zealand, where jade was already well established, but gold deposits had just been discovered (meaning a mining rush was on its way), the prominent local Māori leader Te Otatu emphasized this innately personal connection: 'let the gold be worked by the white men. It was not a thing known to our ancestors. My only treasure is the jade.'[2]

In Chinese, Central American and certain Pacific cultures, jade has been revered for thousands of years. Other highly valued gemstones have typically seen their cultural relevance spread far beyond the places where they were first found and admired – propelled by trade, diplomacy and war – but jade is different. Since the Stone Age it has been fashioned and favoured in its cultural heartlands, in each place prized above all other materials by a loyal following. While that lustre has not dimmed more than 8,000 years later, generally speaking the appreciation of jade has not travelled well.

Instead, its esteem runs narrow and deep in the places that have always considered jade to be the superior stone. Around it has been constructed an inspired symbolism that is one of the most intricate in the history of gemstones, making the sometimes inscrutable jade a stone that must be studied to appreciate how profound the cultural journey of a gem can be – intertwined with a society's understanding of the most fundamental issues of humanity: how to live long, how to live and die well, and how to prepare for what may follow.

Just as its meanings are multi-layered, so too is the material itself. For gemmologists today, everything that is said and understood about jade actually refers to two distinct stones, found in various places, at different and far distant points in history. In addition, some of what has historically been referred to as jade, it would turn out, is nothing of the sort.

THE CATCH-ALL Chinese word for jade – *yù*, a precious stone of great beauty – is one of the oldest characters in the Chinese language and has been in use for at least 5,000 years.[3] It is also a suitably ambiguous label for a stone that has appeared in many forms, not all of which meet the threshold of gemmological jade. These ersatz jades include a mishmash of tough, translucent-to-opaque greenish materials such as serpentine, bowenite, amazonite, marble, quartzite and chrysoprase. They were often grouped together by archaeologists (no doubt to avoid the difficulties of gemmological classification) as 'social' or 'archaeological' jades: similar stones that bear the same symbolic significance as jade, but whose chemistry and mineralogy are quite removed.[4]

What is classified today as 'true jade' comes in only one of two forms: nephrite or jadeite. The former, which comprises almost all Chinese jade until the eighteenth century AD, is an amphibole silicate with a microcrystalline structure, whose tiny interwoven crystals appear bound together in a rope-like manner, giving a fibrous appearance. By contrast, jadeite, a silicate in the pyroxene family, is also composed of minute interlocking crystals, but they are clearly distinct from one another, giving a more granular feel. For much of their history, both nephrite and jadeite have been carved as ornamental hardstones, but in their finest qualities they have also formed beautiful jewels.

It is this texture that has set jade apart from other gem materials, making it more appreciated for its soft, smooth feel, rather than the sharp transparency so expected in other gemstones. It is also this crystalline structure which has stood nephrite and jadeite in good stead over the centuries and across cultures: the result in both cases being an extremely high durability, meaning materials that, while being immensely hard to polish, can withstand bashing and breaking. These were the perfect stones for deadly neolithic clubs and heavenly objects associated with the afterlife.

While both share a tight crystalline structure, the differences are apparent. Jadeite, which has a far richer colour palette and may be found in lavender, orange, black and white, as well as the characteristic green, has a glassier lustre, not unlike that of a microcrystalline quartz, such as chalcedony. In the highest qualities, green and lavender jadeite make up all the most valued jade jewels on the auction market. Nephrite, in its shades of dark spotted green to milky white, has an appearance often described as 'greasy', a soft sheen that led its creamy white examples to be christened 'mutton fat' jade. These were the most highly prized jades of ancient China and the Mughals.

The name 'jade' is actually a Mesoamerican legacy, adapted from the sixteenth-century Spanish term *'piedra de ijada'*. When the *conquistadores* noticed the Aztecs wearing the stone about their waists, and took it to be a cure for pains in the flank and kidney ailments, they named it 'the kidney stone.'* In Latin, it translated as '*lapis nephriticus*', from whence the name 'nephrite' was born.

W HEN I think of jade, I think of China, Hong Kong, and my many stays in the region. For eight years I visited regularly on business, gaining an insight into a jewellery culture a world away from Western expectations and a luxury buyers' demographic where jade was ubiquitous. On my second visit – bringing Princess Margaret's collection for preview, which no doubt contributed to several of the top lots selling to the Far East – I found myself headed to the Jade Market of Yau Ma Tei in a rare moment of free time. A world away from the glitzy Convention Centre of auctions and trade shows in Wanchai, this was a street-level block full of stands stuffed with beads, bangles, and a million different jade look-a-likes from dyed aggregates to plastic and glass.

I'm not sure if I found any true jade on that day, but if I had, I would never have known for certain: a stone notoriously difficult to identify, and subject to a host of different treatments and imitations, even the experts I know now will often send their 'jade' for testing by an experienced laboratory before assuming

---

* It has also been argued that a similar etymology from the Portuguese '*pedra de la mijada*' (literally 'a stone to urinate') was adopted via the Chinese, making this another contender for the root of the word 'jade'. The Chinese also held the belief that jade could help expel kidney stones: one of many apparently coincidental similarities between these ancient jade cultures.

anything. It was a clear lesson in the localized nature of the material – as a Westerner, it was not something I was expected to recognize.

Jade also brings to my mind the extraordinary Mayan burial masks I have seen in the National Museum of Mexico, and the New Zealand *hei-tiki* pendant hidden away in my mother's jewellery box. But above all, it takes me back to China: the first-century royal jade burial suits of the Han dynasty in the Beijing National Museum, and the enormous (over 2 metres high) eighteenth-century sculpture of Emperor Yu Controlling the Flood – the largest in the country – carved in the reign of the Qianlong Emperor and housed in the Forbidden City.

Any understanding of why and how jade matters begins here, where it holds a unique and magical place in common culture. Its usage throughout Chinese history stretches back to at least 6000 BC, when the Xinglongwa culture were making jewellery from jade and using it as part of burial rituals – with jade hoops for the ears and jade rings placed over the eyes, the beginning of a long tradition in both China and Mesoamerica that would enshrine jade's position as a connection and conduit between this life and the next.

During subsequent millennia jade started to be fashioned into symbolic objects: jade figurines from the Honghsan culture (*c.*4700–2900 BC) depict numerous animal forms, from the 'pig dragon' hybrid that was an important burial ornament to more recognizable dragons, eagles that may have represented a con-nection to heaven and the soul's upward journey in the afterlife, and turtles that could similarly have pointed to the underworld.[5] After the Hongshan came the Liangzhu (*c.*3300–2300 BC), whose jade signature was the combination of a *bi* (a circular disc with a hole through the middle) and *cong* (a square tube with a hollowed, circular interior). Widely found in burials, the exact meaning of these objects is unclear, although the symbolism may have been to do with stargazing: a forerunner to the later proverbial Chinese description of astronomers 'looking at heaven through a tube'.[6]

Retelling a story from around this time, an early compendium of Chinese myth and folklore, *The Classic of Mountains and Seas*, makes clear that jade's amuletic value – as a protector from bad luck and evil spirits – was established early on. In reference to a legendary Chinese ruler of the 2600s BC, it notes: 'the Yellow Emperor took jade flowers from the mountains and planted them on the southern side of the Zhong Mountains . . . Since the ghosts and gods fed on jades, by [wearing] them, gentlemen could be protected from ominous things.'[7]

Even in these earliest jade objects, we can discern some of the unique characteristics that would come to define this gemstone throughout its history, across the disparate civilizations that prized it. One was its durability, crucial to the earliest inhabitants across Eurasia, whose tools and weapons such as jade spearheads and axes survive from not only Neolithic China, but also Europe and Russia, where Alpine and Siberian jade provided local sources.*[8] A second characteristic was its cradle-to-grave associations with life itself, from being a symbol of fertility to a gemstone buried with the dead – a fashion that would continue and become increasingly elaborate. And a third was that jade symbolized protection

---

* At the same time as the Xinglongwa burials were taking place in China, jade from the Italian Swiss Alps was being shaped into axe-heads and used all over Europe, some found as far afield as Scotland. On the shores of Lake Baikal in Russia, one Bronze Age burial was found with a couple holding hands, a huge jade dagger beside the woman, and the man with a ring of jade over one eye, very much in the manner of the Xinglongwa inhumation.

from the forces that would do us harm, something that continues today in the form of jade bangles that are routinely put around the wrists of babies in China and prized by their adult owners.* These qualities and associations have endured and deepened with age, while in time new branches of symbolism would also grow, developing into even more refined and nuanced doctrines.

For the vast majority of China's long history with jade, nephrite was predominant. Emerging from the northern Kunlun mountains in Western China, it was found alluvially and mined directly in major centres including Hetian (modern Khotan, in Xinjang province), with rivers known by the Uyghurs as the Yurungkash (White Jade River) and Karakash (Black Jade River).[9] In the seventeenth century one recorded local custom was for women to be sent out to gather jade from the river at night and naked: since jade was masculine (yin) and formed by moonlight, it could best be attracted by the feminine yang of the young ladies.[10] The creamy white and green gems that emerged were harder than the jadeite that was already being found elsewhere, but would not enter Chinese consciousness until the eighteenth century, when it began to arrive in great quantities from Burma.

When jade is mined, it may emerge in anything from small pebbles to massive boulders – one of the largest ever excavated, in 2016, was 19 feet long and weighed around 175 tonnes.[11] The challenge is knowing how much jade one of these rocks may

---

* I remember one client coming into our office in Hong Kong delighted that the valuable jade bangle she had worn and treasured for many years had shattered, because it had taken the hit of the bad luck headed her way. This protection was the reason for wearing it, and to her, worth more than the jewel itself.

contain: jade boulders are often covered with a layer of rough 'skin', and although the outward colour gives a clue, jade may or may not run throughout the interior or be evenly distributed. Although 'windows' can be made in an attempt to spy into the stone, these are still no guarantee of what hides within, and what prize lies (or not) beneath the surface.

Just how much hinges on the gamble that is jade is captured by the ancient story of a legendary Chinese stone that emerged from the state of Chu (modern Hubei and Hunan) in around the eighth century BC. A man named Mr He discovered a boulder of what he believed was such high-quality jade that he immediately brought it to the king. But the royal jeweller declared: 'It is only a stone.' For the crime of having sought to defraud the king, Mr He was condemned to have a foot amputated. When the king died, and his son acceded to the throne, Mr He returned with his jade only to receive another negative verdict and the punishment of having his other foot cut off. After he was said to have wept tears of blood in his distress, an emissary was sent to question him, and was told: 'I do not grieve because my feet have been cut off. I grieve because a precious jewel is dubbed a mere stone, and a man of integrity is called a deceiver.' His find was finally re-evaluated, polished up until its true worth was revealed, and cut into a priceless *bi* disc.[12] Centuries later, this jewel would supposedly be at the centre of a battle between two warring Chinese states, one of whose rulers offered the wealth of fifteen cities for the object (before reneging on the deal) – the origin of the Chinese saying that jade is 'worth many cities'.[13]

Once the jade has been extracted, working with it becomes the next challenge. Nephrite's tightly crystalline structure makes it a difficult stone to craft even with modern tools, which renders the highly developed Neolithic jade trade all the more remarkable.

Hand-cranked rotary machines with abrasives were being used from around 3500 BC, and a reconstruction of traditional techniques estimated that cutting a jade pebble into a flat disc would have been the work of over sixty-eight hours.[14] By the Chinese Bronze Age, vessels were being hollowed out, and inlays of bronze, turquoise, or contrasting colours of jade were being used.[15]

As tools and techniques progressed, the story of jade in China was evolving in other ways. By the age of Confucius (c.551–479 BC), late in what is known as the Spring and Autumn period, a unique development occurred, not seen in the history of other gem materials. There was an increasing emphasis on a new dimension of jade – as a philosopher's stone, a gem that did not just contain value but conveyed virtues and carried strong ethical associations.

In describing jade, Confucius identified eleven *de* (virtues) that epitomized core aspects of his abiding ethical philosophy of decent human behaviour: acting with kindness, respecting one's elders, and the importance of ritual. Each element of jade corresponded with a human virtue – its polish stood for purity, its hardness suggested intelligence, its colour represented loyalty, etc. – thus making the material itself a moral guide.[16] Men were encouraged to wear jade pendants to influence their actions. As Confucius wrote in his *Book of Odes*, 'when I think of a wise man, he seems to be like jade'.[17] The Confucian mantle was taken up six centuries later by the politician and scholar Xu Shen, who associated jade with just five virtues: the warmth of its lustre standing for kindness, its translucent appearance for honesty, its 'tranquil' tone for wisdom, its durability for bravery, and its sharp edges 'not intended for violence' for integrity.*

---

* A discordant final credit, given jade's prominence as a material used in practical as well as decorative weaponry.

In the hands of Chinese philosophers, jade had transcended the traditional matrix for determining the value of a gemstone – its beauty, rarity and durability. When asked why jade was so richly valued and soapstone not at all, Confucius replied that it was not due so much to the comparative rarity of jade, but the stone was so esteemed because of its symbolic value. Jade had, in other words, taken on a life and meaning beyond the physical qualities of its material. It could impart the lessons of how to live as well as acting as an actual bridge to the afterlife. The 'Stone of Heaven', as it has become known, had uniquely developed a didactic function as much as a decorative one. It was the ultimate manifestation of the characteristics associated with a civilized existence and a life well lived: something utterly invaluable.

CONFUCIUS AND those who followed him had elevated jade, but in doing so they had not diverged from its fundamental and long-standing symbolic tradition. The notion that jade has perennially been the stone that represents life, death, the afterlife, and everything in between, reached its zenith during the Han dynasty (220 BC–AD 220), with the creation of some of the most unusual and remarkable gemmological objects ever discovered: full burial suits crafted from several thousand plates of jade.

The use of jade as a burial item went back millennia. It had begun with the interment of ritual jade objects and developed during the Zhou dynasty (1046–256 BC) into jade face masks, constructed from forty to eighty pieces of jade sewn together onto cloth, often incorporating animal imagery. As the art historian Wu Hung has suggested, such objects merged the belief in jade's protective power with the idea that body and soul could find new form in the afterlife, with death considered a process first of preservation and then transformation. '[Because] of the extraordinary hardness and beauty of jade, this natural material was bestowed with magical power to protect or transform the dead . . . particular jade artefacts were created at this time to transform the corpse into an awesome, supernatural being.'[18]

The *yu yi* – burial suits – of the Han took this concept even further. In the most elaborate of them, the jade is shaped to represent facial features – fashioned into the form of the nose and ears. There are jade fingers, realistic anatomical curves, and for Liu Sheng, Prince of Zhongshan in the second century BC, even a set of genitals 'to preserve the prince's sexuality and generativity'.[19] These astonishing creations were thought to be mythical – recorded in the literature but as yet undiscovered – until Liu's tomb was excavated in 1968. And the suit itself was only one part of a comprehensive jade assemblage within the burial. The body underneath had already been 'sealed' with jade plugs, one for every single orifice of the body, and overlaid with jade *bi* discs. The plugs represented the base layer of protection for the body and soul from the ravages of death. Without them, as the first-century philosopher Wang Chong argued, the 'spirit' of a person would escape the body like rice spilling out of the hole in a sack.[20] This idea of sealing the body was furthered by lining the coffin itself with jade.

The function of jade in this context was both physical and spiritual: preserving the body from decay as much as holding in the spirit. It was also about transforming the body into something greater, facilitating the transition from earth to heaven and equipping it for what was believed to come next.[21] In this way, the jade burial suits of the Han dynasty were not only astonishing products of material craftsmanship. They also represented an expression of belief in the power of this gemstone to protect, sustain and reshape life, even after death.

SUCH BELIEFS were not a singular Chinese phenomenon. They also formed the basis of jade's appreciation by the civilizations of Mesoamerica, from the Olmec (*c.*1400–400 BC) who occupied the Gulf of Mexico, to the Maya who spanned parts of modern Mexico, Honduras, Guatemala and El Salvador, and the peoples who made up the eventual Aztec Empire in central Mexico (1428–1521). For the Mesoamericans, jade served a wide variety of ritualistic and symbolic purposes, spanning status and protection, as well as fertility and cosmology.

For the Olmec, who sourced their bright bluish-green jadeite from the Motagua River Valley in Guatemala,* one of the fundamental jade objects was the celt – a rounded axe-head used as a farming tool. Some were inscribed with cosmological imagery – the Mesoamerican version of jade as the stone of heaven and earth.[22] Many were engraved with the image of the Olmec maize deity, and would be planted in the ground to promote the growth of crops. Long jade perforators – whose function it was to let blood from the ears, tongue or genitals – were also used in Olmec and Mayan fertility practices, in offerings made for both crops

---

* The ancient sources of this unusual bluish-green jadeite were long lost until they were rediscovered in the 1970s by American archaeologists Jay and Mary Lou Ridinger. It was this discovery that put an end to the 'Jade Question' – of whether the Mesoamerican cultures were in fact getting their material somehow from Asia, and whether this explained many of the similarities of jade usage between the two.

and the well-being of the community. Self-mutilation through bloodletting was a power-play especially for the elite, a public reminder of their communion with the gods, also indicated by the numbers of perforators found in high-ranking Olmec graves.[23]

Elsewhere jade was used in high-society status symbols, notably in headdresses, ear piercings (including large ear flares) and pectoral ornaments, but also in dental modifications. Mayan burials attest to the practice of drilling and inserting small jade roundels into the front teeth, most likely for simple cosmetic reasons – but possibly also for protective, tribal, religious or intimidatory purposes. Jade was, more straightforwardly, the first choice for rulers across a broad swathe of Mesoamerican history. 'In Classic Maya art', the historian Karl Taube has written, 'jade is so inextricably linked to images of Maya rulers that it is difficult to conceive of them without this precious stone', also suggesting that one way in which potentates established dominance over conquered rivals was to remove their jade jewellery from them.[24]

Jade remained a preferred royal adornment right up to the end of the Aztec Empire. Bernal Díaz del Castillo, who was with Cortés during his conquest of the Aztecs, described how Montezuma II arrived at one of his negotiations with the Spanish, carried on a litter topped with 'a canopy of exceedingly great value, decorated with green feathers, gold, silver, chalchihuis [jade] stones, and pearls'.[25]

One notable aspect of the Mesoamerican craze for jade was how closely its symbolism and cultural references overlapped with the ideas of life, death and the afterlife in China. The connection that the breath of life and the soul were transmitted through the ears and mouth as facilitated by jade was prominent in both cultures. Numerous fantastic Mayan jade burial masks share similarities with the Chinese jade burial suits. The vivid jade death mask of the Maya ruler Pakal the Great, who reigned for sixty-eight years in the seventh century, with perforators poking through ear ornaments and obsidian eyes glaring out, is just one extraordinary example.* Bartolomé de las Casas, a Spanish Dominican friar, provided some insight into the beliefs that underpinned this practice, with reference to the Poqomam, a Maya people spanning Guatemala and El Salvador. 'When it appears that some lord is dying, they had ready a precious stone which they placed at his mouth when he appeared to expire, in which they believe that they took the spirit, and on expiring, they very lightly rubbed his face with it. It takes the breath, soul or spirit.'[26] For the Maya, this was a holistic concept: jade associated with the wind that blew rain onto the life-giving maize fields, and with capturing the final breath of a person that represented their soul.

The fertility associations touching on birth as well as the life-giving necessity of the harvest are summed up by a mythical story surrounding Huitzilihuitl, King of Tenochtitlán (the centre of modern-day Mexico City and one of the three city states that would soon unite to become the Aztec Empire). The king had sought a diplomatic marriage alliance with a neighbouring ruler

---

* I was lucky to see this magnificent object in Mexico in 2012, given its modern-day backstory. On Christmas Eve 1984 thieves broke into the National Museum of Anthropology and stole over 100 priceless artefacts coming from Mexican excavations, including Pakal's jade mask. It was only four and a half years later that the items were recovered, from the home of one of the two veterinary school drop-outs who had carried out the robbery; for a year it had been stashed in a suitcase on top of a closet. The pair were finally turned in to the police by drug traffickers with whom they had tried to make a deal.

but been laughed out of court, his messengers sent away and told not to return. His counterpart, ruler of the cotton-producing Cuernavaca region, openly mocked any suitor who he said could only afford to clothe his daughter in the traditional agave plant fibres. But the King of Tenochtitlán was not deterred. Rather than retreating to lick his wounds, he consulted his deity and, so the folkloric story goes, took a more direct approach. 'He stood within the boundaries of the lord of Cuernavaca. Then he shot a dart, a prettily painted and marvellously crafted reed, in the centre of which was inserted a precious jade – most valuable and shimmering brightly.' His aim was true and the gem landed almost in the lap of the girl whose hand in marriage he had sought. Entranced by the gemstone, she put it in her mouth and accidentally swallowed it. She fell pregnant, they married after all, and their child, Montezuma I, became the second Aztec Emperor (1440–66).[27]

Framing jade as a gem of royalty, fertility and destiny, the tale neatly captures what made this gemstone such a long-standing cultural icon across the contrasting civilizations that prized it. These included the Māori, Eastern Polynesian settlers who arrived in New Zealand in the second half of the thirteenth century, a people entirely reliant on neolithic technologies. One of their most important finds was nephrite from the South Island. Together with a similar dark green silicate called bowenite, they called this 'greenstone' *pounamu*. With it they could hunt, fish, build, and carve canoes and spears. Yet the stone was symbolic as well as practical, summed up by the use of jade to manufacture two significant Māori objects: the *hei-tiki* (neck ornaments) and *mere* (paddle-shaped clubs).

The former were highly stylized flattened carved jade figurines worn as necklaces that held links back to one's ancestors and may have represented Tiki, the first man in Māori mythology, or a fertility goddess.[28] One was brought back by Captain Cook after his first voyage and presented to King George III in 1771; he would subsequently return with European-made versions to offer in return.*[29] The *mere* was both a practical weapon, designed for clubbing and slicing, and an important ritual object, a status symbol that would be passed down through generations. Its power – *mana* – was believed to increase with the succession of owners and the finest examples were owned by chiefs and accorded supernatural properties. One ruler's *mere* 'was said to be invisible to his enemies, and to hide itself and reappear at his call'.[30] It was also deemed an honour to be killed using such a sacred weapon: captured leaders were known to surrender their own *mere* so that it could be used to strike the fatal blow.

For these three contrasting cultures in which jade was a common thread, the meaning and majesty of the gem was deep-rooted and ancient. Yet its history was not static, and the story of jade continued to find new ways to unfold. In particular, new discoveries and prominent patrons would help to refresh the relevance of jade, ensuring it sustained its cultural dominance into the modern world.

IN ALL the long history of Chinese jade adoration and appreciation, one figure stands out. A long-reigning ruler who

---

* This anticipated an ironic turn of events in the early twentieth century, as tourism to New Zealand took off and a market developed in selling jade objects featuring Māori designs. With the local industry unable to keep pace with demand, *pounamu* would be exported to the European cutting centre of Idar Oberstein in Germany, manufactured into 'Māori' objects, and re-exported for sale to the tourist market in New Zealand.

obsessively penned hundreds of poems about jade and owned tens of thousands of jade objects, it is probably no exaggeration to say that the Qianlong Emperor (r.1735–1796) was the most committed jade connoisseur of all time. As well as studying and writing about the stone, he was known for playing an active role in the design and production of jade objects by his court craftsmen, going into painstaking levels of detail and inspecting both designs and raw materials before allowing work to proceed. He had old-fashioned aesthetic preferences: he scolded makers who indulged in over-elaborate designs that deviated from traditional models, and even requested that antique signatures be baked into some objects to fake their age.[31]

The emperor's jade cravings were more than a personal interest: they also provided a medium through which to connect the achievements of his reign, especially in territorial conquest, to familiar Chinese symbols and beliefs. After the Qing had seized what is now Xinjiang from the Mongols in 1761, the emperor was encouraged by his nephew to commission a set of jade chimes to mark the victory, which had brought a major jade-producing region back under Chinese rule. By commissioning the jade chimes – a Confucian symbol of harmony and good governance – with stone from the territory he had just conquered, the Qianlong Emperor was underlining his legitimacy as a ruler, his connection to sacred Chinese beliefs, and the righteousness of his military exploits.[32]

As well as keeping a tight rein on his own craftsmen, the Qianlong Emperor coveted the jades that were being manufactured elsewhere, notably in Mughal India. They were the subject of over seventy of his jade poems, one of which noted, perhaps ruefully, that 'although Khotan produces both raw and carved jade, all the best carvings are from Hindustan'.[33] By the time of his reign, the Mughal expertise in jade carving was well established, having been nurtured under the rule of Jahangir and then Shah Jahan in the previous century.

The best Mughal jades were creamy-white 'mutton fat' jades, originating from Hetian in Xinjiang, but carved with incomparable finesse for the Mughal courts. Some cups were so fine that they were practically translucent, the light passing through their edges like sheer white porcelain.* It was jades like this to which the emperor was referring in his poem, 'In Praise of a Hindustani Drinking Vessel', when he wrote of 'water mills grinding the jade as thin as paper'.[34]

Such was the quality of the Mughal work and the Qianlong Emperor's admiration for it that a two-way trade became established in his reign, by which Chinese jade would be exported to India for manufacture, and the products re-imported; he also urged his own craftsmen to imitate the styles that were emerging from the Indian workshops. Mughal decadence dictated that many jade objects had a banqueting function, from wine goblets to serving bowls, although in common with other jade-admiring cultures, Mughal craftsmen also fashioned weaponry, including hilts and sheaths for swords or daggers, and even archers' rings.[†35]

---

* One such piece is the wine cup of Shah Jahan, made in 1657, and now in the Victoria and Albert Museum. Carved from a single piece of white jade, it has a foot in the form of a lotus flower and an asymmetric gourd-shaped body which curls into a ram's head terminal, elegantly twisting around. The sides are so thin and look so fragile that it is almost impossible to comprehend that white jade is one of the toughest natural materials in the world.

† These were often highly decorated, inset with cabochon emeralds and rubies, and sometimes also diamonds, in the typical Mughal style also used for jade jewels.

Equally, jade was already established as a material of symbolic importance to the Timurids, the Mughals' cherished ancestors, who had imbued its associations with life, death and the afterlife. In 1425 the Timurid Emperor Ulugh Beg obtained a huge jade slab which he set over the burial site of his grandfather Timur (Tamerlane) in Samarkand.[36]

The Qianlong Emperor's jade legacy would be significant: he had not only been a proactive and demanding patron of China's jade industry, but the military campaigns of his reign had brought two major jade-producing regions under the banner of the Qing dynasty: the mountains of Khotan to the west, and to the south, the mines of the Kachin Hills in the northern region of modern Myanmar. The second of these was of massive importance, as it marked the introduction to China of jadeite jade – the vivid green gem known as *fei cui* (named after the bright iridescent feathers of the kingfisher). This exceptional Burmese jadeite would cause a sensation and cast nephrite – the staple of Chinese jade from which all the Qianlong Emperor's signature pieces had been crafted – into the shade.

The import of jadeite into China began in his reign, but it would not be until the nineteenth century, and the influence of another royal jade fanatic, that so-called 'imperial jade' became part of the gem's long-running story in China. This was Empress Dowager Cixi (r.1861–1908), one of the most powerful women in the history of Imperial China, *de-facto* ruler of the Qing dynasty in its twilight years. For Cixi, jade was both a personal favourite, with supposedly enough jewels to fill 3,000 sandalwood boxes, and a symbol of her authority: the imperial seal she was presented on her sixtieth birthday was cut from jade, with an elaborate handle comprising two interlocking dragons.[37] She had rings of jade, jade objects decorating her bedroom, jade chopsticks to eat with, she applied her lipstick with a jade hair-pin, and she drank her tea in the gardens of the Summer Palace from a jade cup, flavoured with rose, honeysuckle and jasmine flowers that would be brought to her in a jade bowl.[38]

The Empress Dowager was not just a collector but a market-maker. Her other great gemstone obsession, pink tourmaline from a mine in Pala, San Diego, became so sought after in nineteenth-century China as a result of her influence, that it created a boom in the Californian tourmaline industry. Over a century after her death, Chinese demand for this relatively unheralded gemstone remains.[39] Cixi had similar pulling power when it came to jade, helping to cement the popularity of vivid green jadeite objects, such as the pearl-fringed ring she wore for an American artist, Katharine Carl, to paint her portrait in 1904.[40] Her final flourish in jade came with her burial, where the Empress Dowager was entombed alongside a panoply of white and green objects resembling those she had so prized in life. Carved jade butterflies decorated her hair, watermelons rested by her feet, alongside both white- and green-peeled melons, a lotus leaf, ten peaches, two cabbages and twenty-seven miniature Buddhas, jadeites all.*[41]

Late Imperial China was not the only place where jade was flourishing in the late nineteenth and early twentieth century. A major discovery of Siberian nephrite in 1826, with another major find in 1851, became part of the Fabergé repertoire of home-grown

---

* The mausoleum suffered a devastating event in 1928, when a military-style operation under the direction of a Chinese warlord ransacked the tomb. The Empress's precious jades were stolen – it was reported that the butterflies were ripped so violently from her head that her hair was torn from her scalp – and although the robbers were well-known, they bribed their way out of any retribution and the matter was never resolved.

gems produced for the Tsars, used to carve a variety of boxes and vessels, as well as featuring on several of the imperial eggs.[42]

It was with Fabergé – who had opened their first European shop in London in 1903 to better serve such patrons as King Edward VII and Queen Alexandra[43] – that jade began to make its presence felt on the Western jewellery scene. By the 1920s a fashion for orientalizing jewels saw jadeite appearing in jewellery and vanity cases by the likes of Cartier, displaying striking, contrasting colour combinations – of green, red, black – that were emblematic of the movement. Jade became part of the stock of Western jewellers who were looking to imitate Eastern styles and colours, without ever coming close to matching the status that the gem held in China and elsewhere.

In America, it was one of the richest women in the world who took the lead in buying into the oriental aesthetic and who was one of the keenest collectors of jade in the West. That was Barbara Hutton, the original Million Dollar Baby, granddaughter of the retail tycoon Frank Whitfield Woolworth. On her twenty-first birthday, she gained access to a trust fund worth just shy of $50 million (the near equivalent to $1 billion today). Yet she would live a troubled existence despite her vast wealth: her mother died by suicide when she was four years old, and Hutton married and divorced on seven occasions, the third time to Cary Grant (a union the press dubbed 'Cash and Cary'). 'I suppose it is incomprehensible to the average person that a girl who was as beautiful as Barbara could have inherited a sense of inadequacy along with forty-five million dollars,' was the arch verdict of the gossip columnist Elisa Maxwell.[44]

Amid all this difficulty, one thing Hutton could love without complication was jewellery. Already obsessed with jade from an early age, she would visit the Asian specialist store Gump's as a young girl to learn what to look for in the stone and how to spot a counterfeit.[45] She followed the reports of the looting of Cixi's tomb in 1928.[46] On her trip to Shanghai in 1934 she purchased a jadeite bangle that had once belonged to Empress Cixi herself.[47]

Her most celebrated jadeite jewel, however, was a record-breaker of epic proportions: a spectacular necklace of twenty-seven enormous, delicious, translucent, and matching vivid green jadeite beads. Presented on the occasion of her first marriage in 1933, the jade beads were sent to Cartier to be mounted on a simple diamond clasp, and, a year later, on a more striking circular ruby and diamond closure.

This necklace first came up for auction in 1988, setting a world record and selling for $2 million; when it reappeared on the auction block six years later in 1994 it doubled that record, achieving $4.2 million; and finally, at its last auction appearance in 2014, it broke the record once more, this time realizing an incredible $27 million, making it easily the most expensive jadeite jewel to go under the hammer.[48] The buyer was Cartier, who acquired it for their own historic Cartier Collection.

When it came up on public view in 2014, one look was enough to confirm that this was a truly stunning world-class jewel. One piece of jadeite of such an even and intense translucent green is rare on its own, but to have so many large beads (between 15.4 and 19.2 millimetres), perfectly matched in colour and size, and of such fine texture and translucency, was simply remarkable. This was a level of quality that would be difficult to surpass, and rumours circulated at the time of the sales that the jade had all come from a single famous boulder, and that the beads were originally worn in eighteenth century Imperial China.[49]

Holding this necklace took me a step closer to understanding a gemstone that can often feel like it lives in a world of its own, whose meaning and importance is not intuitive but needs to be learned. But looking at these jadeites, like a string of juicy grapes or crunchy apples all in a line, it was impossible not to get it. There was something about the vivid green of the colour that drew me in and made immediate sense of the jade's long-standing connection to life itself. You can see why this is a gemstone people might fall in love with and never want to stop collecting; why Empress Cixi used to carry little jadeite cabbages around just to play with.[50] It is vibrant and vivacious – not so much shining as almost appearing to come alive.

Like a window being polished onto a newly excavated nephrite boulder, it had taken me one step closer to the gemmological wonder and cultural phenomenon that is jade: a material that has been so much more than a gemstone for the civilizations that have revered it. Jade is extraordinary not just for its aesthetic qualities, but for the comprehensive utility it has shown across 8,000 years in human hands, serving almost every purpose you could imagine. Jade has been functional: a tool for farming and fighting. It has been ceremonial: part of rituals from bloodletting to burial. It has been a decorative marker of status, from hairpins to whole body suits, and it has been amuletic in its perceived ability to offer power and protection. And, uniquely among gemstones, it has been adopted as a vessel for ethics and morality.

The story of jade is a true gemstone epic, one that has spoken to some of the deepest human concerns, and achieved totemic status in some of humanity's most influential civilizations. Other gems have travelled further and had more universal relevance than jade, but none has gone so far in inspiring such deep cultural allegiance, or provoking so much thought about the nature of our lives on earth, and what it means to live them to the fullest.

Afterword

O F ALL the great privileges a life in the gem business has afforded me – from seeing the first glint of a rough gemstone emerging from a mine or riverbed, to holding some of the most magnificent examples humanity has known in my hand – one factor above all keeps me coming back for more. It is the way this industry is defined by a combination of uncertainty and discovery, something that keeps it as fresh to me now as it was when I first knocked on the doors of Bond Street jewellers at the start of my career.

Panning the earth of a Sri Lankan river, you never know which handful is going to yield a pebble that will be polished on the jeweller's wheel into a sparkling sapphire. Travelling the world, your ears are always open for the first rumour that something interesting has been found – something that might become the next mega gem deposit, making plentiful a stone that was previously rare, or even bringing an entirely new variety blinking into the light of day. Gem-testing in the laboratory, your eyes are always peeled for the newest treatment or an unusual inclusion under the microscope. In the auction room, you wait for the gem that knocked your socks off during viewing to come up for sale, to see what it will make, and if it will be the one to set a new world record. While being shown around collectors' houses and jewellers' vaults, you will often be handed anonymous-looking boxes that may contain miracles: secret family heirlooms, legendary stones that have gone into hiding, or simply stunning examples that remind even a seasoned gem connoisseur of why they love these objects and dedicate their lives to understanding them.

These known unknowns mean it is impossible to become jaded by the world of gemstones, one that encompasses millennia of history yet is also constantly changing and evolving – carried forward on a tide of new discoveries, changing fashions and economic fluctuations. In the course of my career so far, I have seen shifts that would have surprised me in my days as an auction-house trainee. Already pricey rubies have seen their values soar still higher, to levels that would then have been difficult to imagine. Esoteric brown- and grey-coloured diamonds, almost unmarketable then, have become desirable now. While the much-maligned spinel is showing sure signs that its reputation is recovering from the damage that was done to it by the ruby promoters and synthetic producers of the late nineteenth and early twentieth centuries.

These ups and downs are part of the story of every gemstone, and have been throughout their history. I could not put it better than the Persian scholar al-Biruni, who wrote in the eleventh century: 'The prices of jewels are not stable. There is no law governing their prices, and there is no reason why these prices should not fluctuate with time and place. Each country, each nation carries its own temper. Furthermore, at one time nobles begin to sell them off and at others, to stock them. Stones are plentiful at one time and scarce at another. God grants honour to some and disgrace to others.'[1]

Gemstones are the earth's creation, but they are a human fascination and as such at the behest of human whims and urges: from how new discoveries tip the scales of supply and demand to how war and diplomacy can disrupt trade routes and marketing may shape tastes and trends.

Yet this constant business of ebb and flow barely leaves a mark when we look at the history of gemstones in its totality: how they have blazed a trail of aesthetic, cultural and financial importance across human civilization. Whatever the reasons a particular stone may be prized more or less in a particular place and at a point in time, none have eroded the fundamental human love of gems and our desire to attach both value and meaning to them. Since the first bright stone was picked up on the side of a riverbed thousands of years ago, we have coveted, gifted, traded, studied and fought over these precious pebbles. Humans have admired gemstones for the unique way in which they interact with light to create a magical aesthetic experience, for the deep psychological resonance of the colours they contain, and for what they represent as an object to be pursued – whether as miner, trader, jeweller or connoisseur.

While our knowledge about gemstones has expanded beyond measure, our attitude towards them, and the feelings they evoke, have not really changed. These are objects that expose continuities in the human experience – transcending generations, cultures and even civilizations. One of the most striking discoveries of my career has been the extent to which our reasons for prizing gems today overlap with those recorded in texts hundreds or even thousands of years old. Teaching gemmology, one of the first questions I would ask my students about a stone is what colour they associate it with. The answers are almost always the same as they would have been at any point in history.

Another important continuity is the global footprint of the gemstone, and how these objects move around the world as they pass from mine to marketplace, workshop, production centre and finally the auction house or jewellery boutique. Today that may mean that an emerald mined in Zambia (where commercial production began in the 1970s) could go to Dubai or Singapore to be sold, Tel Aviv or New York for cutting and treating, and then on to Hong Kong, Geneva or London to enter the consumer market. Nor does the story end there, in a never-ending cycle where gems might be recut, retreated and resold at any place and any time. These sprawling supply chains are no modern creation. As with the ancient world's garnet trade, gemstones have been criss-crossing the world for millennia: a portable source of wealth and prestige whose travels provide a window into trade routes and manufacturing hubs, then and now. Excavations of tools and stones have often provided historians with an archaeological trail of minute pieces of evidence, allowing them to reconstruct the shape of a trade more sophisticated and interconnected than we often allow.

The trade in which I have spent my career is comprised of many moving parts and contrasting faces, some of which reflect its status as a sophisticated modern industry, and others which would be recognizable to gem hunters and jewellers of any era. Gemstones still get scooped up from riverbeds, plucked from open-pit mines and heat-treated under charcoal fires. In other places they are the subject of heavy industry, with machinery that specializes in crushing kimberlite ore or pumping the water in vast quantities out of an emerald mine. There are many types of mining just as there is a huge variety of gem materials, an extensive range of different cutting techniques, and a variety of industries within the industry – from the highly centralized model De Beers created in the diamond business to the more traditional approach that exists elsewhere, with many, widely distributed links in the chain as a stone passes from source to end buyer.

It can seem convoluted, complicated and inconsistent, yet

this heterogeneity is the source more of fascination than frustration. I know that every mine I visit is likely to be at least a little different from any other I have seen. In every jeweller's workshop I will learn or encounter something new. And most importantly, every time a gemstone crosses my desk to be assessed or valued, I am seeing something that is unique – the product of its own journey with its own distinctive features, forged by a combination of trace elements, exposure to heat and pressure, and a journey through human hands that will never exactly be replicated. Every stone tells its own story, and so many of those stories remain to be told. The future in gemmology is always every bit as exciting as its extensive history.

As a result, one of the incentives for working in this world is the knowledge that your journey through it is limitless. There will always be something new around the corner – new gems, new technologies – and there is so much that remains unseen and unknown. The need to learn and experience as much as possible has driven my desire for discovery: from working in the auction business to setting up and running a gemmology school and curating a museum collection. I have wanted not only to see and understand as many parts of the business as possible, but to experience them for myself, hands on. I have handled as many gemstones as I possibly can, tried on every tiara I was allowed to, and gone underground at every mine that would admit me. In the same way I have travelled to new places at every opportunity, visiting archaeological sites and museums all over the world as much as mines and workshops: to understand how gems have made their progress from out of the ground and into our hands. As much as possible, I have tried to follow the journey of the gems themselves.

In doing so, I have had the good fortune to meet the people who make the industry work, holding up the sometimes unsteady supply chain from miner to designer. These people are the true gem geniuses when it comes to following the geological trail of new discoveries, sensing where mining activity should be focused, making a stone sing on the cutting table, and being able to arrive at an almost instant valuation when a new specimen is put in front of them. We do them a disservice if we treat the story of gemstones simply as one of expensive jewellery and famous collectors. That is merely the icing on the cake, of an industry whose heart is in the mines, the workshops and the markets of the gemstone world.

In that universe, one of the undoubted stars is Sri Lanka, the Island of Gems itself, and a place I have always considered my gemstone home from home. It has a mine-to-market model where the spirit of gems is at the heart of much of the industry there. I want to leave the last word to a mine manager I met on one of my many visits. We were stuck underground together after a small incident that prevented anyone from going up or down for a short time. We got talking and I learned a bit more about the mine and his life operating it – the cooperative system resulting in the excitement, pride and relief at discovery. I told him of my appreciation at being allowed in to visit his place of work, and see somewhere that held the fate of so many individuals and whole families in its hands. I knew I would be back and asked what if anything I could bring next time. 'Just bring me more people,' he said. 'We are proud of what we do and want the world to know.'

This book is for him, and the many men and women I have met who have dedicated their lives in search of a prize that is born hidden from human eyes, yet once revealed is almost incomparably beautiful, valuable and precious. I cannot take the world to him, but perhaps, in some of the stories in this book, I can bring to the world the secrets behind some of the most incredible gems ever discovered.

Acknowledgements

THIS BOOK HAS been a long time in the making, spanning nearly twenty-five years of marvellous moments in an extraordinary industry. I will forever be thankful to those who shared this world with me from the beginning, especially Kieran McCarthy at Wartski (who opened the first jewellery safe to me on Bond Street many moons ago and still discusses Fabergé, and more, with me today), Keith Penton, who introduced me to the auction industry from the outset, and my other colleagues at Christie's – especially Raymond Sancroft-Baker and David Warren – whose knowledge and sense of humour inspired many memorable moments. I am indebted to Francois Curiel who gave me the opportunity of a lifetime in working on the Princess Margaret sale, and also to Lord Snowdon.

Early on, there were a couple of very special people who helped set me out on this path in the first place: Martin Henig, who introduced me to and mentored me on ancient gems, and David Miller, whose inspirational teaching of Classics has ultimately resulted in my ability to read and translate first-hand the ancient Latin and Greek lapidaries, a useful skill indeed.

Many gem experts and historians have helped me with research, advice and brainstorming. My thanks in particular to: Jack Ogden, Ken Scarratt and Eddie Cleveland for helping me unravel some of the spinel story, Laurent Cartier, Annemarie

ACKNOWLEDGEMENTS

Jordan Gschwend and Martin Travis for their expertise on pearls, Olivier Baroin for his knowledge of Suzanne Belperron, Bernhard Berger for the story of the Patiala ruby choker, Bruce Bridges for sharing his family history, Mark Cullinan for his take on the Cullinan Diamond, and Richard Hughes for all his inspiring gem research. A fun thanks to David Shara and Josh Cohn at Optimum diamonds for showing me the Baby Blue Hope and regularly teasing me about my reaction.

Other illustrious friends and colleagues who have given me input, advice and much-appreciated support include Vivienne Becker, Beatriz Chadour-Sampson, Richard Edgcumbe, Charlotte Gere and Amanda Triossi. Amongst these are two very special *grands remerciements* to Cynthia Unninayer, who kindly proofread the first full manuscript, and Jeffery Bergman: both kept me going with their kind encouragement as much as their intellectual input.

I put together the greater part of *Precious* during my first year as Senior Curator at the V&A, juggling a new environment and diverse responsibilities. The two in tandem could not have been possible without the unerring support of my kind and brilliant Keeper, James Robinson, our Director, Tristram Hunt, and a department full of very empathetic colleagues. I'm grateful to our entire Metalwork and DAS team for their understanding and patience, and my close team of Jessica Rosenthal Mcgrath, Clare Phillips and Judith Crouch in particular, for much-needed moments of help and kindness.

Other museum and collections colleagues who shared ideas and information include Sue Strong at the V&A, Sue Brunning at the British Museum, Hazel Forsyth at the Museum of London (who had also shown me the Cheapside Hoard in person), Caroline de Guitaut at the Royal Collection Trust, Violette Petite and

César Imbert at the Cartier Archives and Pascale Lepeu at the Cartier Collection.

I owe huge thanks to all those who have taken me on their journeys as well as those who have come on mine, including: Vincent Pardieu (my first field trip to Thailand with the original field gemmologist himself), Hpone Phyo Kan Nyunt and Kyaw Thu (my Burmese guides with a sense of humour as large as their gemmological knowledge), Keith Barron (whose Rock Creek sapphires will always be close to my heart) and, above all, Armil, Sarrah and Savi Sammoon (my Sri Lankan friends and family, and the heart of all my stays in their country). The kindness and help Dinesh offered in helping me trace my grandfather's steps in Sri Lanka will never be forgotten, and we all miss him greatly. Also to those who have come on my field trips – students who have become friends and taught me as much about the world as the trips themselves, including Eurosia (my colleague and right-hand), Joanne, Angie, Elena, Peter and many more. And to all my students from my teaching days in Geneva, Zurich, Lucerne, Hong Kong and Shanghai – I know that you came to my classes, but you all taught me more than you will ever know.

I am enormously grateful to everyone at Transworld who has made this dream a reality: the wonderful Katherine Cowdrey who helped obtain all our beautiful images and more, Milly Reid, Hannah Winter, Richard Mason, Holly McElroy, Barbara Thompson, Katrina Whone, Phil Lord, Bobby Birchall and Irene Martinez. And, above all, Susanna Wadeson, the best editor I could ever have hoped for, whose brilliance, kindness, experience and understanding kept me excited, focused and sane. A special thanks to my agent, Adam Gauntlett, who came to me with the idea of *Precious*, and persuaded me into this exciting project.

To my great friends who have been there at the right time,

with various forms of help and encouragement, humour, food and photocopies: Andrew, Anthea, Duncan, Genevieve, Nadine and James, Parthiv, Philip, Pedro, Ralph, Wyndham and my very dear friend Stéphanie. I am especially grateful to Charles and Angela and all my friends at Rousham for the many happy summer hours spent writing there.

But most of all I want to thank my parents, who put me up and put up with me through several intense months of working and reworking drafts back home. They allowed me to babble and brainstorm about bling incessantly, proofread every version of every chapter, and let me bore them senseless with minutiae without ever showing it. Instead, they offered constant encouragement, as they have always done. None of this could have happened without them. And lastly (and least), to Dudu dog, who kept me entertained and distracted just when I needed it most.

Picture Acknowledgements

Chapter motifs:

Tony and Hannah at Global Creative Learning

Images:

César Imbert and Violette Petite at Cartier archives

Dirk Weber and Marius Winzeler at the Dresden State Art
Collections

Jeffrey Post and Adam Mansur at the Smithsonian

Vincent Vanhoucke and Jessica Wyndham at Sotheby's

Alexandra Damianos and Rahul Kadakia at Christie's

Nicola Griffin and Emily Barber at Bonhams

Josephine Smart and Elliott Graff at Graff Diamonds

Federica Boido and Lauren Chase Weaver at Harry Winston

Graeme Thompson and Larry West at L. J. West Diamonds

Barbara Berkowitz and Mitch Erzinger at the Elizabeth Taylor
Trust and Archive

Sarah Davis and Lee Siegelson at Siegelson

Jack Glover Gunn at the Victoria and Albert Museum

Gerard Panczer at the University of Lyon

Pui Visoot and Vlad Yavorskyy at IVY/Yavorskyy

Geoffrey Munn

Notes

1 Emerald

1 H. Forsyth, *London's Lost Jewels: The Cheapside Hoard*. London: Philip Wilson Publishers, 2013.

2 R. Weldon and C. Jonathan, 'The Museum of London's Extraordinary Cheapside Hoard', *Gems & Gemology*, vol. 49 (3), 2013, pp. 126–37.

3 G. Giuliani et al., 'Oxygen Isotopes and Emerald Trade Routes Since Antiquity', *Science*, vol. 287, 2000, pp. 631–3.

4 K. Lane, *Colour of Paradise: The Emerald in the Age of Gunpowder Empires*. New Haven, CT: Yale University Press, 2010, p. 10.

5 J. Gosling, 'The Cheapside Hoard Confusion', *The Journal of Gemmology*, vol. 24 (6), April 1995, pp. 395–400.

6 Forsyth, *London's Lost Jewels*.

7 Personal comment by Hazel Forsyth: the Muzo origin was identified by the Colombian emerald expert, Ron Ringsrud.

8 J. Sinkankas, *Emerald and Other Beryls*. Prescott, AZ: Geoscience Press, 1989, p. 408; S.H. Ball, 'Historical Notes on Gem Mining', *Economic Geology*, vol. 26, 1931, pp. 681–738.

9 Lane, *Colour of Paradise*, pp. 50–8.

10 B.G. Gunn, *The Instruction of Ptah-Hotep and The Instruction of Ke'gemni: The Oldest Books in the World*. London: John Murray, 1906, p. 42.

11 Sinkankas, *Emerald and Other Beryls*, p. 5; S. García-Dils De La Vega et al., 'The Emerald Mines of Wadi-Sikait (Egypt) from a Diachronic Perspective: Results of the 2020 and 2021 Seasons of the Sikait Project', *Trabajos de Egiptología, Papers on Ancient Egypt*, no. 12 (2021), p. 20.

12 Lucan, *Pharsalia*, X, 139–40 (trans. A.S. Kline); see also J. Ogden, 'Cleopatra's Emerald Mines: The Marketing of a Myth', *The Journal of Gemmology*, vol. 38 (2), June 2022, pp. 156–170, on how Cleopatra's association with emeralds was romantically exaggerated in the nineteenth century.

13 Pliny the Elder, *Natural History* (trans. D.E. Eichholz), XXXVII, 62.

14 Ibid, 76 (author's translation).

15 H.M. Said, *Al-Beruni's Book on Mineralogy: The Book Most Comprehensive in Knowledge on Precious Stones*. Islamabad: Pakistan Hijra Council, 1898, p. 146.

16 Seneca, *Epistles*, 90, 33; W. Jensen, *The Leyden and Stockholm Papyri: Graeco-Egyptian Chemical Documents from the Early 4th Century* AD. Cincinnati, OH: University of Cincinnati, 2008.

17 For example, Victoria and Albert Museum, museum nos. 629-1884 (variscite) and 8853-1863 (variscite and amazonite).

18 M. Henig and H. Molesworth, *The Complete Content Cameos*. Turnhout, Belgium: Brepols, 2018, pp. 172–3.

19 S.P. Scott, *The Civil Law*, vol. XV, Cincinnati, OH: The Central Trust Company, 1932, Book XI, Title 11.1.

20 Sinkankas, *Emerald and Other Beryls*, p. 100.

21 Various versions of this myth exist; the one here was related by the Colombian emerald expert Guillermo Galvis.

22 Sinkankas, *Emerald and Other Beryls*, p. 30.

23 Lane, *Colour of Paradise*, p. 102.

24 Christie's, 'A Magnificent Emerald, Gold and Enamel Wine Cup', *Arts of India*, 24 September 2003, Lot 168.

25 Christie's, 'An Emerald and Diamond Pendant Brooch, by Bulgari', *The Collection of Elizabeth Taylor: The Legendary Jewels, Evening Sale (I)*, 13 December 2011, Lot 29.

26 G.F. Kunz, *The Magic of Jewels and Charms*. Philadelphia, PA: J.B. Lippincott Company, 1915, p. 135.

27 G.F. Kunz, *The Curious Lore of Precious Stones*. Philadelphia, PA: J.B. Lippincott Company, 1913, pp. 380–2.

28 Theophrastus, *On Stones* (trans. E.R. Caley and J.F.C. Richards), 24.

29 Pliny the Elder, *Natural History*, XXXVII, 64.

30   H.M. Sherman, 'The Green Operating Room at St Luke's Hospital',
     California State Journal of Medicine, vol. 12 (5), May 1914, pp. 181–3.

31   Pliny the Elder, *Natural History*, XXXVII, 63.

2   Ruby

1   C.M. Enriquez, 'Fire-Hearted Pebbles from Burma', Asia Magazine, vol.
     30 (10), 1930, pp. 722–5, reproduced at www.palagems.com/burma-ruby.

2   R. Hughes, *Ruby and Sapphire: A Gemologist's Guide*. Bangkok: RWH
     Publishing/Lotus Publishing, 2017, p. 520.

3   C.O. Alley et al., 'Apollo 11 Laser Ranging Retro-Reflector: Initial
     Measurements from the McDonald Observatory', *Science*, vol. 167
     (3917), 1970, pp. 368–70.

4   A. Sasso, 'The Geology of Rubies', *Discover Magazine*, 25 November
     2004.

5   *The Garuda Purana*, chs. 68–72.

6   Hughes, *Ruby and Sapphire*, p. 356.

7   G.F. Kunz, *The Curious Lore of Precious Stones*. Philadelphia, PA: J.B.
     Lippincott Company, 1913, p. 103, citing Taw Sein Ko, communication
     from his 'Burmese Necromancy'.

8   M. Krzemnicki, 'Pigeon Blood Red and Royal Blue: Working Towards an
     International Standard', *GAHK Seminar*, March 2017, p. 31.

9   J.B. Nelson, 'Colour Filters and Gemmological Colorimetry', *Journal of
     Gemmology*, vol. XIX (7), 1985, p. 621.

10   T.H. Hendley, *Indian Jewellery*. London: Journal of Indian Art,
     1909, p. 33.

11   *The Travels of Sir John Mandeville*. London: Macmillan and Co,
     1915, p. 131.

12   Kunz, *The Curious Lore of Precious Stones*, p. 256.

13   E.W. Streeter, *Precious Stones and Gems: Their History, Sources and
     Characteristics*. London, 1892, pp. 162–3.

14   Theophrastus, *On Stones* (author's translation), 18.

15   J.-B. Tavernier, *Travels in India*, vol. 2, trans. V. Ball. Cambridge:
     Cambridge University Press, 2012, p. 101.

16 Robert Gordon, 'On the Ruby Mines near Mogok, Burma', *Proceedings of the Royal Geographical Society and Monthly Record of Geography*, vol. 10 (5), 1888, pp. 261–75.

17 Hughes, *Ruby and Sapphire*, p. 513.

18 H. Nadelhoffer, *Cartier: Jewelers Extraordinary*. London: Thames & Hudson, 1999, p. 159.

19 F. Cartier Brickell, *The Cartiers*. New York: Ballantine Books, 2019, p. 342.

20 S. Stronge, 'Indian Jewellery and the West: Stylistic Exchanges 1750–1930', *South Asian Studies*, vol. 6 (1), 1990, pp. 143–55.

21 Cartier Brickell, *The Cartiers*. p. 161.

22 Personal comment by Bernhard Berger, Head of Cartier Tradition at the time, who discovered the bracelet.

23 Christie's, 'The Patiala Ruby Choker', *Maharajahs and Mughal Magnificence*, 19 June 2019, lot 272.

3   Sapphire

1 A. Lucas et al., 'Sri Lanka: Expedition to the Island of Jewels', *Gems & Gemology*, vol. 50 (3), 2014, pp. 174–201; H. Molesworth, 'Ancient Sapphires and Adventures in Ceylon', *Gems & Jewellery*, 26 (3), 2017, pp. 28–31; P. Zwaan, 'Sri Lanka: The Gem Island', *Gems & Gemology*, vol. 18 (2), 1982, pp. 62–71.

2 A. Lang, *The Arabian Nights Entertainment*. London: Longmans Green and Co, 1898, p. 177.

3 E. Gass, 'One Nation, Many Gemstones', *Gems & Jewellery*, 29 (1), 2020, p. 20.

4 S. Saeseaw, W. Vertriest and A. Palke, 'Sapphires from Anakie, Australia: A Closer Look at Blue, Yellow, Green and Bicolour Sapphires', *GIA*, 23 August 2021.

5 T. LaTouche, 'The Sapphire Mines of Kashmir', *Records of the Geological Survey of India*, vol. XXIII, Calcutta (1890), pp. 59–60.

6 G. Panczer et al., 'The Talisman of Charlemagne: New Historical and Gemological Discoveries', *Gems & Gemology*, vol. 55 (1), 2019, pp. 30–46.

7 Exodus 26:4; 24:10–12.

8 K. Link, 'Age Determination of Zircon Inclusions in Faceted Sapphires', *The Journal of Gemmology*, vol. 34 (8), 2015, pp. 692–700; W. Xu et al., 'Age Determination of Zircon Inclusion in Kashmir Sapphire with U-Pb Dating', *Journal of Gems and Gemmology*, vol. 22 (1), 2020, pp. 1–12.

9 R. Edgcumbe, 'My Beautiful Sapphires: Queen Victoria's Sapphire and Diamond Coronet by Kitching and Abud', in *Liber Amicorum in Honour of Diana Scarisbrick, A Life In Jewels*, Ad Ilissum (2022), pp. 172–88.

10 C. Gere, 'Love and Art: Queen Victoria's Personal Jewellery', *Victoria & Albert: Art & Love*. London: Royal Collection Trust, 2012, p. 7.

11 L. Field, *The Queen's Jewels: The Personal Collection of Elizabeth II*, New York: Harry N. Abrams, Inc., 1997, p. 149.

12 ABC News, 14 November 2010.

4 Garnet

1 S. Brunning, *The Sword in Early Medieval Northern Europe: Experience, Identity, Representation*, Woodbridge: Boydell Press, 2019, pp. 89–90.

2 Personal comment.

3 N. Adams, 'Reading the Sutton Hoo Purse Lid'; C. Hicks, 'The Birds on the Sutton Hoo Purse', *Anglo-Saxon England*, vol. 15. Cambridge: Cambridge University Press, 1986, p. 160; R. Allen, 'A Stag Stands on Ceremony: Evaluating Some of the Sutton Hoo Finds', *Bulletin of the John Rylands Library*, vol. 79 (3), 1997, pp. 167–76.

4 L. Thorensen and K. Schmetzer, 'Greek, Etruscan and Roman Garnets in the Antiquities Collection of the J. Paul Getty Museum', *The Journal of Gemmology*, vol. 33 (7–8), 2013, p. 201.

5 Theophrastus, *On Stones* (trans. E.R. Caley and J.F.C. Richards), 18.

6 Pliny The Elder, *Natural History* (trans D.E. Eichholz), XXXVII, 91–2.

7 R. Blake and H. De Vis, *Epiphanius De Gemmis: The Old Georgian Version and The Fragments of The Armenian Version and The Copto-Sahidic Fragments*, London: Christophers, 1934, p. 199.

8 British Museum, museum no. EA37532.

9 In the Altes Museum, Berlin.

10 J. Spier, 'A Group of Ptolemaic Engraved Garnets', *The Journal of the Walters Art Gallery*, vol. 47, 1989, pp. 21–38.

11 N. Adams, 'The Garnet Millennium: The Role of Seal Stones in Garnet Studies', *Gems of Heaven* (eds C. Entwistle and N. Adams), London: The British Museum, 2011, pp. 10–24.

12 N. Adams, 'Late Antique, Migration Period and Early Byzantine Garnet Cloisonné Ornaments: Origins, Styles and Workshop Production', PhD Thesis, University College, London, 1991, p. 62; K. Schmetzer et al., 'The Linkage Between Garnets Found in India at the Arikamedu Archaeological Site and their Source at the Garibpet Deposit', *The Journal of Gemmology*, vol. 35, no. 7, 2017, pp. 598–627.

13 Sotheby's, *Old Master Sculpture and Works of Art*, 3 December 2014, lot 12.

14 N. Adams, ' "Carbunculus ardens: The Garnet on the Narses Cross in Context" ', Dumbarton Oaks Papers 69, Dumbarton Oaks Research Library and Collection, Washington DC, 2015, p. 154.

15 F. Dell'Acqua, 'The Carbunculus (Red Garnet) and the Double Nature of Christ in the Early Medieval West', *Konsthistorisk tidskrift/Journal of Art History*, vol. 86 (3), 2017, pp. 158–72.

16 H. Hamerow, 'The Circulation of Garnets in the North Sea Zone ca. 400–700', *Gemstones in the First Millennium AD: Mines, Trade, Workshops and Symbolism* (eds A Hilgner, S. Greiff, D. Quast), Mainz, 2013, p. 77; Patrick Périn et al., 'Provenancing Merovingian Garnets by PIXE and μ-Raman Spectrometry, *Millennium Studies*, vol. 5 (1), 2007, pp. 69–75.

17 J. Ljungkvist, J.S. Lundahl, P. Fröhlund, 'Two Workshops with Garnet Crafts in Gamla Uppsala, *Gemstones in the First Millenium AD*', p. 97; C. Pion, B. Gratuze, P. Périn and T. Calligaro, *Bead and Garnet Trade between the Merovingian, Mediterranean, and Indian Worlds*, 2020, p. 23.

18 D. Stehlíková, 'Bohemian Garnet Jewellery: 1,500 Years of Czech Treasures', *Jewellery History Today*, vol. 28, Winter 2017, pp. 3–5.

19 K. Schmetzer and H.-J. Bernhardt, 'Garnets from Madagascar with a Colour-Change of Blue-Green to Purple', *Gems & Gemology*, Winter 1999, pp. 196–201.

20 C. Schwarzinger, 'Determination of Demantoid Garnet Origin by Chemical Fingerprinting', *Monatsh Chem*, vol. 150, 2019, pp. 907–12.

21 A. K. Snowman, *Carl Fabergé, Goldsmith to the Imperial Court of Russia*. London: Debrett's Peerage Limited, 1979, p. 111.

22 C. Bridges, 'The History of Tsavorite', via www.tsavorite.com.

23 C. Unninayer, 'Campbell Bridges Savoring Tsavorite', *The International Jewelry Quorum*, September/October 2007, pp. 4–5; L. Vigne and

E. Martin, 'Discovering Tsavorite, Kenya's Own Gemstone: An Interview with Campbell Bridges', *Kenya Past and Present*, Kenya Museum Society, October 2008, pp. 61–8.

24 Delicate hand-mining of one of the world's most valuable green gems – https://www.youtube.com/watch?v=yrpoYDkUNVA.

25 Vigne and Martin, 'Discovering Tsavorite', p. 64.

26 tiffany-ad-campbell-bridges-1974-new-yorker.pdf (tsavorite.com).

27 B. Branstrator, '50 Years of Tsavorite: A Q&A with Bruce Bridges', nationaljeweler.com; H. Molesworth, 'The Evolution of Pricing of Gems and Jewellery', *InColor*, Spring/Summer 2020, pp. 51–3; Bruce Bridges personal comment.

5 Pearl

1 P. Morand (trans. E. Cameron), *The Allure of Chanel*. London: Pushkin Press, 2013, p. 144.

2 E.M. Sehasseh et al., 'Early Middle Stone Age Personal Ornaments from Bizmoune Cave, Essaouira, Morocco, *Science Advances*, vol. 7 (39), 2021, eabi8620.

3 V. Charpentier, C.S. Phillips and S. Méry, 'Pearl Fishing in the Ancient World: 7500 BP', *Arabian Archaeology and Epigraphy*, 23 (2012), pp. 1–6.

4 Nonnus, *Dionysiaca*, XXXII.26, in W.H.D. Rouse, *Nonnus: Dionysiaca*, Cambridge: Loeb Classical Library, 1940–2, pp. 446–7.

5 P. Southgate and J. Lucas (eds), *The Pearl Oyster*. Oxford: Elsevier Science, 2008, pp. 2–3.

6 G.F. Kunz and C.H. Stevenson, *The Book of the Pearl: The History, Art, Science and Industry of the Queen of Gems*. New York: The Century Co, 1908, p. 37.

7 Pliny the Elder, *Natural History* (trans. D.E. Eichholz), IX, 54.

8 'The True Story Behind How Pearls Are Made', Smithsonian Voices, National Museum of Natural History, *Smithsonian Magazine*, 5 August 2021.

9 J. Taylor and E. Strack, 'Pearl Production', in Southgate and Lucas (eds), *The Pearl Oyster*, pp. 274–6.

10 S. Akamatsu, L.T. Zansheng, T.M. Moses, K. Scarratt, 'The Current Status of Chinese Freshwater Cultured Pearls', *Gems & Gemology*, vol. 37 (22), 2001, p. 96.

11 K.D. Ericson, 'Nature's Helper: Mikimoto Kokichi and the Place of Cultivation in the Twentieth Century's Pearl Empires', PhD thesis, Princeton University, 2015, p. 1.

12 F. Cartier Brickell, *The Cartiers*. New York: Ballantine Books, 2019, p. 336.

13 K. Scarratt and S. Karampelas, 'Pearl Evolution, Pearl Testing and the Historic Use of X-Rays', *InColor*, Spring/Summer 2020, pp. 82–6.

14 P. Craddock (ed.), *Scientific Investigation of Copies, Fakes and Forgeries*. Oxford: Butterworth-Heinemann, 2011, p. 403; W. Schumann, *Gemstones of the World*, 4th edn. New York: Sterling, 2009, p. 264; A. Farn, *Pearls: Natural, Cultured and Imitation*. London: Butterworth-Heinemann, 2013, pp. 118–19.

15 Revelation 21:21.

16 Pliny the Elder, *Natural History* (author's translation), IX, 54, and XXXVII, 6.

17 'Life of Julius Caesar', ch. 47 in C. Suetonius Tranquillus, *The Lives of the Twelve Caesars*. London: Loeb Classical Library, 1913, p. 65.

18 M. Deutsch, 'Caesar and the Pearls of Britain', *The Classical Journal*, vol. 19 (8), 1924, pp. 503–5.

19 Pliny the Elder, Natural History, XXXIII, 47.

20 'Life of Julius Caesar', ch. 43 in Suetonius, *The Lives of the Twelve Caesars*, p. 61.

21 'Erythraean Pearls in the Roman World: Features and Aspects of Luxury Consumption (Late Second Century BCE–Second Century CE)', hal.science, pp. 14–15.

22 B. Ullman, 'Cleopatra's Pearls', *The Classical Journal*, vol. 52 (5), 1957, pp. 193–201.

23 R. Hibbard, 'Pearls: Piety, Poetry and Pre-Raphaelites – Part One', *V&A Blog*, 1 October 2013.

24 B. Chadour-Sampson and H. Bari, *Pearls*. London: V&A Publishing, 2013, pp. 50–2.

25 M. Bailey, 'Salvatrix Mundi: Representing Queen Elizabeth I as a Christ Type', *Studies in Iconography*, vol. 29, 2008, pp. 175–215; P. McClure and R.H. Wells, 'Elizabeth I as a Second Virgin Mary', *Renaissance Studies*, vol. 4 (1), 1990, pp. 38–70.

26 M. Perri, ' "Ruined and Lost": Spanish Destruction of the Pearl Coast in the Early Sixteenth Century', *Environment and History*, vol. 15 (2), 2009, pp. 129–61.

27 A. Belsey and C. Belsey, 'Icons of Divinity: Portraits of Elizabeth I', in L. Gent and N. Llewellyn (eds), *Renaissance Bodies: The Human Figure in English Culture, c. 1540–1660*. London: Reaktion Books, 1990, p. 13; also L. Montrose, *The Subject of Elizabeth: Authority, Gender, and Representation*. Chicago, IL: University of Chicago, 2006, p. 147.

28 D. Macmillan, 'The Power behind the Pearl', *Vides*, vol. 7, 2019, pp. 239–52.

29 Chadour-Sampson and Bari, *Pearls*, pp. 72–4.

30 G. Bapst, *Histoire des Joyaux de la Couronne de France*. Paris: Librairie Hachette, 1889, pp. 22–3; D. Scarisbrick, *Tudor and Jacobean Jewellery*. London: Tate Publishing, 1995, p. 17.

31 E. Taylor (R. Peltason, ed.), *Elizabeth Taylor: My Love Affair with Jewelry*. New York: Simon & Schuster, 2002, p. 82.

32 Lord F. Hamilton, *Here, There and Everywhere*. New York: George H. Doran, 1921, p. 167.

33 For the best historical account of the pearl, see A. Jordan Gschwend, *La Peregrina, An Extraordinary Renaissance Pearl*, London: Paul Holberton, 2023, where the author also translates '*peregrina*' in sixteenth-century Spanish as meaning 'rare' or 'unique'.

34 Hamilton, *Here, There and Everywhere*, p. 167.

35 E. Taylor (R. Peltason, ed.), *Elizabeth Taylor: My Love Affair With Jewelry*, p. 84.

36 Ibid, p. 91.

37 V. Becker, *Mikimoto: The Pearl Necklace*. New York: Assouline, 2016, p. 25.

38 Personal comment by Oriole Cullen, Senior Curator of Textiles and Fashion at the Victoria and Albert Museum.

39 *Chanel: A Woman of Her Own*. New York: Henry Holt and Co, 1990, p. 197.

40 A. Sato and L. Cartier, 'The Value of Pearls: A Historical Review and Current Trends', *Gemguide*, May/June 2022, p. 4; R. Shor, 'From Single Source to Global Free Market: The Transformation of the Cultured Pearl Industry', *Gems & Gemology*, vol. 43 (3), Fall 2007, p. 201.

41 F. Shen, *Pearl: Nature's Perfect Gem*. London: Reaktion Books, 2022, p. 120.

6 Spinel

1   R. Hughes, *Ruby and Sapphire*. London: RWH Publishing, 1997, pp. 280–2.

2   All references to the tenuous history of the Black Prince's Ruby from J. Ogden, 'The Black Prince's Ruby: Investigating the Legend', *The Journal of Gemmology*, vol. 37 (4), 2020, pp. 360–72.

3   Personal comment by the gemmologist Ken Scarratt, who has examined the spinel out of its mount.

4   C. Blair, S. Bury, A. Grimwade, R.R. Harding, E.A. Jobbins, D. King, R.W. Lightbown and K. Scarratt, *The Crown Jewels: The History of the Coronation Regalia in the Jewel House of the Tower of London*. London: The Stationery Office, 1998, vol. II, pp. 55–7.

5   R. Hughes, 'The Rubies and Spinels of Afghanistan: A Brief History', *The Journal of Gemmology*, vol. 24 (4), 1994, p. 257.

6   al-Biruni, in R. Hughes, *Ruby and Sapphire: A Collector's Guide*, Bangkok: Gem and Jewelry Institute of Thailand, 2014, p. 194.

7   J. Princep in the *Journal of the Asiatic Society of Benghal*, 1932, in Hughes, 'The Rubies and Spinels of Afghanistan', p. 258.

8   S. Dale, *Babur: Timurid Prince and Mughal Emperor, 1483–1530*. Cambridge: Cambridge University Press, 2018, pp. 100–18.

9   D.M. Dirlam et al., 'Gemstones in the Era of the Taj Mahal and the Mughals', *Gems & Gemology*, vol. 55 (3), Fall 2019, pp. 294–319.

10 The A'in-I Akbari (Institutes of Akbar).

11 S. Stronge, 'The Myth of the Timur Ruby', *Jewellery Studies*, vol. 7, 1996, p. 5.

12 A.S. Melikian-Chirvani, 'The Red Stones of Light in Iranian Culture. I. Spinels', *Bulletin of the Asia Institute*, vol. 15, 2001, pp. 77–110.

13 Stronge, 'The Myth of the Timur Ruby', p. 5.

14 M.H. Fisher, *A Short History of the Mughal Empire*. London: I.B. Tauris, 2016, p. 167.

15 Queen Victoria, *Journals*, vol. 32. Windsor Castle, 1851, pp. 164–5.

16 Stronge, 'The Myth of the Timur Ruby', pp. 8–9.

17 E.W. Streeter, *Precious Stones and Gems: Their History, Sources and Characteristics*. London, 1892, p. vii.

18 J.M. Ogden, 'Gem Knowledge in the Thirteenth Century: The St Albans Jewels', *The Journal of Gemmology*, vol. 37 (8), 2021, p. 823.

19 Ibid, p. 209.

20 *Financial News*, London, 28 February 1889, in Hughes, *Ruby & Sapphire*, p. 317.

21 Streeter, *Precious Stones and Gems*, p. 212.

22 R. Webster, *Gems: Their Sources, Descriptions and Identification, Vol 1*. London: Butterworths, 1962, pp. 303–13.

23 Personal comment by Jeffery Bergman.

24 V. Pardieu, 'Hunting for "Jedi" Spinels in Mogok', *Gems & Gemology*, vol. 50 (1), Spring 2014, p. 46.

25 M.S. Krzemnicki et al., 'Cobalt-Bearing Blue Spinel from Lukande near Mahenge, Tanzania', *The Journal of Gemmology*, vol. 38 (5), 2023, pp. 474–93.

7 Quartz

1 R. Kostov and O. Pelevina, 'Complex Faceted and Other Carnelian Beads from the Varna Chalcolithic Necropolis: Archaeogemmological Analysis', *Geoarcheology and Archeolmineralogy, Proceedings of the International Conference*, Sofia, 2008, pp. 67–72.

2 'Queen Puabi's Headdress', Penn Museum, Museum nos. B11693–4, B11708–12B.

3 V. de Michele, 'The "Libyan Desert Glass" Scarab in Tutankhamen's Pectoral', *Sahara* 10, 1998, pp. 107–10.

4 M. Henig and H. Molesworth, *The Complete Content Cameos*. Turnhout, Belgium: Brepols, 2018; for an overview of these more private jewels, see H. Molesworth and M. Henig, 'Love and Passion: Personal Jewels in Late Antiquity from the Content Collection', *Gems of Heaven*, eds C. Entwistle and N. Adams. London: The British Museum, 2011, pp. 179–85.

5 J.J. Pollini, 'The Gemma Augustea: Ideology, Rhetorical Imagery, and the Creation of a Dynastic Narrative', in P.J. Holliday, *Narrative and Event in Ancient Art*. Cambridge: Cambridge University Press, pp. 258–98; G.M.A. Richter, *Engraved Gems of the Romans: A Supplement to the History of Ancient Art*. London: Phaidon, 1971, p. 104.

6   J.-B. Giard, *Le Grand Cameé de France*. Paris: Bibliothèque Nationale de France, 1998; Richter, *Engraved Gems of the Romans*, pp. 104–5.

7   F.S.K., 'The *Tazza Farnese* Reconsidered', *American Journal of Archaeology*, vol. 96 (2), 1992, pp. 249–54.

8   Pliny the Elder, *Natural History* (author's translation), XXXVII, 121; 124.

9   W.T. Pavitt and K. Pavitt, *The Book of Talismans, Amulets and Zodiacal Gems*. London: William Rider & Son, 1922, pp. 138–9.

10 G.F. Kunz, *The Curious Lore of Precious Stones*. Philadelphia, PA: J.B. Lippincott Company, 1913, p. 267.

11 Pliny the Elder, *Natural History*, XXXVII, 139 and 105.

12 S.H. Ball, 'A Historical Study of Precious Stone Valuations and Prices', *Economic Geology*, vol. 30 (5), 1935, pp. 630–42, via palagems.com/ball-gem-prices.

13 Sotheby's, *The Jewels of the Duchess of Windsor*, Geneva, 2–3 April 1987.

14 'The Uncrowned Jewels', BBC Documentary (1987), via youtube.com

15 P. Corbett, W. Landrigan and N. Landrigan, *Jewelry by Suzanne Belperron*. London: Thames & Hudson, 2015, p. 221.

16 C. Horyn, 'Modern, Before the World Was', *The New York Times*, 19 December 2012.

17 Corbett, *Jewelry by Suzanne Belperron*, p. 114.

18 G. King, *The Duchess of Windsor: The Uncommon Life of Wallis Simpson*. New York: Citadel Press, 2000, p. 403.

19 C. Brown, *Hello Goodbye Hello: A Circle of 101 Remarkable Meetings*. New York: Simon & Schuster Paperbacks, 2012, p. 331.

20 Corbett, *Jewelry by Suzanne Belperron*, p. 227.

21 'The Uncrowned Jewels'.

22 E. Possémé, 'Suzanne Belperron 1900–83', in L. Mouillefarine and E. Possémé, *Art Deco Jewelry: Modernist Masterworks and Their Makers*. London and New York: Thames & Hudson, 2009, pp. 74–6.

23 C. Horyn, 'Modern, Before The World Was', *New York Times*, 19 December 2012.

24 Corbett, *Jewelry by Suzanne Belperron*, p. 128.

25 'The Uncrowned Jewels'.

26 Sotheby's, *The Jewels of the Duchess of Windsor*, April 1987, lots 81 and 31.

8 Diamond

1   B. Gorelik, 'The Cullinan Diamond and its True Story', *Jewellery Studies*, June 2015, p. 6.

2   E. Caesar, 'The Woman Shaking Up the Diamond Industry', *The New Yorker*, 27 January 2020.

3   Gorelik, 'The Cullinan Diamond', p. 9.

4   J. Fishman, *My Darling Clementine: The Story of Lady Churchill*. New York: David McKay Company, Inc, 1963, pp. 31–2.

5   K. Scarratt and R. Shor, 'The Cullinan Diamond Centennial: A History and Gemological Analysis of Cullinans I and II', *Gems & Gemology*, vol. 42 (2), Summer 2006, pp. 120–32.

6   Personal comment by Mark Cullinan.

7   S. Shirey and J. Shigley, 'Recent Advances in Understanding the Geology of Diamonds', *Gems & Gemology*, vol. 49 (4), Winter 2013, pp. 188–222.

8   'How Diamond-Bearing Kimberlites Reach the Surface of Earth: Acidification Provides the Thrust', *Science Daily*, 26 January 2012; J.K. Russell et al., 'Kimberlite Ascent by Assimilation-Fuelled Buoyancy', *Nature*, vol. 481, pp. 352–6.

9   M.B. Kirkley et al., 'Age, Origin and Emplacement of Diamonds: Scientific Advances in the Last Decade', *Gems & Gemology*, vol. 27 (1), Spring 1991, pp. 2–25.

10 A. Janse and P. Sheahan, 'Catalogue of Worldwide Diamond and Kimberlite Occurrences: A Selective and Annotative Approach', *Journal of Geochemical Exploration*, vol. 53 (1–3), 1995, pp. 73–111.

11 K.J. Anderson, 'Diamonds', *MRS Bulletin*, vol. 19 (8), 1994, p. 85.

12 K. Scarratt and R. Shor, 'The Cullinan Diamond Centennial', p. 124.

13 R. Shamasastry, *Kautilya's Arthashastra*. Bangalore: Government Press, 1915, pp. 105, 145, 343.

14 M. Frank, *Flight of the Diamond Smugglers: A Tale of Pigeons, Obsession and Greed Along Coastal South Africa*. London: Icon Books Ltd, 2021, pp. 11–12, 28–30.

15 Caesar, 'The Woman Shaking Up the Diamond Industry'.

16 T. Nicols, *Lapidary: or, The History of Precious Stones*. Cambridge: Thomas Buck, 1652, p. 17.

17   J. Ogden, *Diamonds: An Early History of the King of Gems*. London: Yale University Press, 2018, p. 10.

18   J. Courtney Sullivan, 'How Diamonds Became Forever', *The New York Times*, 3 May 2013.

19   Imperialist and Dr Jamieson, *Cecil Rhodes: A Biography and Appreciation*. London: Chapman & Hall Ltd, 1897, p. 5.

20   S. Reilly, 'De Beers SA: A Diamond is Forever', NYU Stern, Case Number MKT04-01, December 2004.

21   J.L. Ghilani, 'De Beers' "Fighting Diamonds": Recruiting American Consumers in World War II Advertising', *Journal of Communication Enquiry*, vol. 36 (3), 2012, pp. 222–45.

22   Sullivan, 'How Diamonds Became Forever'.

23   E. Epstein, 'Have You Ever Tried to Sell a Diamond?', *The Atlantic*, February 1982.

24   Ibid.

25   U. Friedman, 'How an Ad Campaign Invented the Diamond Engagement Ring', *The Atlantic*, February 2015.

26   Bain & Company, 'The Global Diamond Industry 2018', pp. 24–8.

27   J. Hummel, 'Diamonds Are a Smuggler's Best Friend: Regulation, Economics, and Enforcement in the Global Effort to Curb the Trade in Conflict Diamonds', *The International Lawyer*, vol. 41 (4), 2007, pp. 1145–69.

28   W. Dalrymple and A. Anand, *Koh-i-Noor: The History of the World's Most Infamous Diamond*. London: Bloomsbury, 2018, p. 2.

29   Ibid, p. 12.

30   Ibid, p. 282.

31   Ibid, p. 222.

32   S. Shah, 'Romancing the Stone: Victoria, Albert and the Koh-i-Noor Diamond', *A Journal of Decorative Arts, Design History, and Material Culture*, vol. 24 (1), 2017, p. 38.

33   Ibid, pp. 39–40; E.W. Streeter, *Diamonds*. London: George Bell & Sons, 1895, p. 42.

9 Coloured Diamond

1 References to the history of the Wittelsbach via R. Dröschel, J. Evers and H. Ottomeyer, 'The Wittelsbach Blue', *Gems & Gemology*, vol. 44 (4), Winter 2008, pp. 348–63.

2 'Graff Accused of "Painting over a Rembrandt" by Gemologists', *Daily Telegraph*, 29 January 2010.

3 D. Crossland, 'Billionaire Accused of Vandalism for Recutting Dh60m Wittelsbach Blue', *The National*, 7 February 2010.

4 'Bavarian Crown Jewel Given Controversial Makeover', *Spiegel International*, 28 January 2010.

5 R. Kurin, *Hope Diamond: The Legendary History of a Cursed Gem*. New York: Smithsonian Books, 2006, p. 3.

6 Ibid. p. 22.

7 Ibid, p. 27.

8 J. Ogden, 'Camels, Courts and Financing the French Blue Diamond: Tavernier's Sixth Voyage', *The Journal of Gemmology*, vol. 35 (7), 2017, p. 648.

9 Kurin, *Hope Diamond*, p. 59.

10 F. Farges, S. Sucher, H. Horovitz and J.-M. Fourcault, 'The French Blue and the Hope: New Data from the Discovery of a Historical Lead Cast', *Gems & Gemology*, vol. 45 (1), 2009, p. 4.

11 J. Ogden, 'Out of the Blue: The Hope Diamond in London', *The Journal of Gemmology*, vol. 36 (4), 2018, pp. 318–19.

12 S. Lee, *Dictionary of National Biography*. New York: Macmillan and Co, 1891, p. 329.

13 Ogden, 'Out of the Blue', p. 328.

14 E.W. McLean, *Father Struck It Rich*. Boston, MA: Little, Brown and Company, 1936, p. 155.

15 Ibid, p. 171.

16 F. Farges et al., 'The French Blue and The Hope'; E. Gaillou et al., 'The Wittelsbach–Graff and Hope Diamonds: Not Cut from the Same Rough', *Gems & Gemology*, vol. 46 (2), Summer 2010, pp. 80–8.

17 McLean, *Father Struck It Rich*, p. 174.

18 Kurin, *Hope Diamond*, p. 207.

19 Ibid, p. 61.

20 Farges et al., 'The French Blue and the Hope', pp. 4–5.

21 McLean, *Father Struck It Rich*, p. 177.

22 Ibid, p. 179.

23 Ibid, pp. 174–5.

24 E. Smith et al., 'Blue Boron-Bearing Diamonds From Earth's Lower Mantle', *Nature*, vol. 560, 2018, pp. 84–7.

25 'The Magic of Coloured Diamonds', via www.gia.edu/gia-news-research-worlds-fascination-fancy-colored-diamonds

26 J.E. Shigley et al., 'Discovery and Mining of the Argyle Diamond Deposit, Australia', *Gems & Gemology*, vol. 37 (1), 2001, pp. 26–41.

27 Christie's, *Magnificent Jewels*, 10 November 2015, lot 409; Sotheby's, *Magnificent Jewels and Noble Jewels*, 11 November 2015, lot 513.

28 Sotheby's, *Magnificent Jewels and Noble Jewels*, 13 May 2014, lot 507.

10 Jade

1 B. Diaz del Castillo (trans. J. Lockhart), *The Memoirs of the Conquistador Bernal Diaz Del Castillo*, vol. I. London: J. Hatchard and Son, 1844, p. 278.

2 E. Best, *The Stone Implements of the Māori*. A.R. Shearer (Government Printer), 1974, p. 175.

3 F. Ward, *Jade*. Bethesda, MD: Gem Book Publishers, 1996, p. 24; R. Hughes et al., *Jade: A Gemologist's Guide*. Bangkok and Colorado: Lotus Publishing and RWH Publishing, 2022, p. 17.

4 G.L. Barnes, 'Understanding Chinese Jade in a World Context', *Journal of the British Academy*, vol. 6 (2018), pp. 9–13.

5 B. Lakomska, 'The Meaning of Animal Motifs in Neolithic China Based on Examples of Jade Figurines and Shell Mosaics', *Art of the Orient*, vol. 9 (2020), pp. 7–26; E. Childs-Johnson, 'Jades of the Hongshan Culture: The Dragon and Fertility Cult Worship', *Arts Asiatiques*, vol. 46 (1991), p. 93.

6 E. Childs-Johnson, 'Speculations on the Religious Use and Significance of Jade Cong and Bi of the Liangzhu Culture', Throckmorton Fine Art catalogue, Liangzhu, 2012.

7   Y. Shu-Xian, L. Wan-er, 'Jade Myths and the Formation of Chinese
    Identity', *Journal of Literature and Art Studies*, vol. 7 (4), April 2017, p.
    380.

8   P. Pétrequin et al., 'Alpine Jades in the European Neolithic', *LI Riunione
    Scientifica dell'Istituto Italiano di Preistoria e Protostoria*, Istituto Italiano
    di Preistoria e Protostoria, Forlì, Italy, 2016, pp. 99–107; O.I. Goiunova
    and A.G. Novikov, 'Jade Artefacts from Bronze Age Cemeteries in the
    Cis-Olkhon Area, the Western Coast of Lake Baikal', *Archaeology,
    Ethnology and Anthropology of Eurasia*, 2018, 46 (4), pp. 33–41.

9   Barnes, 'Understanding Chinese Jade in a World Context', p. 2.

10 Hughes et al., 'Jade', p. 76.

11 www.forbes.com/sites/trevornace/2016/10/24/miners-find-massive-jade-
    boulder-worth-170-million/?sh=6cdbbaf44f7d

12 J. Minford and J. Lau (eds), *Classical Chinese Literature: An Anthology of
    Translations, Volume I: From Antiquity to the Tang Dynasty*. New York:
    Columbia University Press and the Chinese University of Hong Kong,
    2000, p. 224.

13 J. Lin, 'What's Behind Jade's Mystical Appeal', *Apollo*, 12 November 2016,
    pp. 72–7.

14 Ibid.

15 M. Wang and G. Shi, 'The Evolution of Chinese Jade Carving
    Craftsmanship', *Gems & Gemology*, vol. 56 (1), p. 38.

16 J. Walker, 'Jade: A Special Gemstone', in *Jade* (R. Keverne, ed.). New
    York: Springer Science + Business Media, 1991, p. 32.

17 Hughes et al., 'Jade', p. 15.

18 W. Hung, *The Art of the Yellow Springs: Understanding Chinese Tombs*.
    London: Reaktion Books, 2010, p. 132.

19 Ibid, p. 134.

20 J. Shi, *Modeling Peace: Royal Tombs and Political Ideology in Early China*.
    New York: Columbia University Press, 2020.

21 Hung, *The Art of the Yellow Springs*, p. 136.

22 K. Taube, 'The Symbolism of Jade in Classic Maya Religion', *Ancient
    Mesoamerica*, vol. 16 (2005), pp. 23–5; K. Taube, *Olmec Art at
    Dumbarton*. Washington, DC: Dumbarton Oaks Research Library and
    Collection, 2004, pp. 122–7.

23 Taube, *Olmec Art*, pp. 122–6.

24 Taube, 'The Symbolism of Jade', p. 23.

25 Diaz del Castillo, *The Memoirs of the Conquistador Bernal Diaz Del Castillo*, vol. I, p. 221.

26 S. Houston, D. Stuart and K. Taube, *The Memory of Bones: Body, Being and Experience among the Classic Maya*. Austin, TX: University of Texas Press, 2006, p. 142.

27 C. Townsend, *Fifth Sun: A New History of the Aztecs*. Oxford: Oxford University Press, 2019, pp. 38–9.

28 D. Austin, *Te Hei Tiki: An Enduring Treasure in a Cultural Continuum*. Wellington: TE Papa Press, 2019, p. 22.

29 Royal Collection Trust, inventory number 69263.

30 A. Hamilton, *The Art Workmanship of the Māori Race in New Zealand*. Wellington: The New Zealand Institute, 1896, p. 185.

31 M. Zhang, *Chinese Jade: Power and Delicacy in a Majestic Art*. San Francisco, CA: Long River Press, 2004, pp. 17–24.

32 Y. Wu, 'Chimes of Empire: The Construction of Jade Instruments and Territory in Eighteenth-Century China', *Late Imperial China*, vol. 40 (1), June 2019, pp. 43–85.

33 R. Skelton, 'Islamic and Mughal Jades' in Keverne (ed.), *Jade*, p. 293.

34 J. Watt, *Chinese Jades from the Collection of the Seattle Art Museum*. Seattle, WA: Seattle Art Museum, 1989, p. 112.

35 S. Markel, 'Carved Jades of the Mughal Period', *Arts of Asia*, vol. 17 (6), November–December 1987, pp. 123–30.

36 O. Grabar, *Islamic Visual Culture, 1100–1800: Constructing the Study of Islamic Art, Volume II*. Aldershot: Ashgate, 2006, pp. 78–80.

37 The Palace Museum, 'Green Jade Seal with Empress Dowager Cixi's Honorary Title and Entwined Dragon Knob', (dpm.org.cn)

38 J. Chang, *Empress Dowager Cixi: The Concubine Who Launched Modern China*. London: Jonathan Cape, 2013, p. 173; M. Yu, *Chinese Jade*. Cambridge: Cambridge University Press, 2011, p. 131.

39 D. Lee, 'Chasing the Glint of a Different-Colored Gem', *Los Angeles Times*, 15 July 2006. https://www.latimes.com/archives/la-xpm-2006-jul-15-fi-chinagems15-story.html

40 The Palace Museum, 'Jadeite Ring', (dpm.org.cn).

41 Yu, *Chinese Jade*, p. 133; A. Levy and C. Scott-Clark, *The Stone of Heaven The Secret History of Imperial Green Jade*. Weidenfeld & Nicolson: London, 2001, p. 208.

42 M.V. Burtseva et al., 'Nephrites of East Siberia: Geochemical Features and Problems of Genesis', *Russian Geology and Geophysics*, vol. 56, 2015, pp. 402–10.

43 K. McCarthy and H. Faurby, *Fabergé: Romance to Revolution*. London: V&A Publishing, 2021, p. 103.

44 E. Maxwell, *I Married the World*. London: William Heinemann, 1955, p. 164.

45 C. D. Heymann, *Poor Little Rich Girl The Life and Legend of Barbara Hutton*. Pocket Books: New York, 1986, p. 24.

46 A. Levy and C. Scott-Clark, *The Stone of Heaven*, p. 234.

47 Ibid. p. 243.

48 Christie's, *Magnificent Jewels*, 12 May 1988, lot 672; Christie's, *Magnificent Jadeite Jewellery*, 31 October 1994, lot 898; Sotheby's, *Magnificent Jewels and Jadeite*, 7 April 2014, lot 1847.

49 Levy and Scott-Clark, *The Stone of Heaven*, p. 294.

50 Yu, *Chinese Jade*, p. 131.

Afterword

1 al-Biruni, *Kitab al-Jamahir fi Ma'rifat al Jawahir*, in Hughes, *Ruby & Sapphire: A Collector's Guide*, Bangkok: Gem and Jewelry Institute of Thailand, 2014, p. 285.

Index

About the Author

HELEN MOLESWORTH'S career has spanned the global gem and jewellery industry, from roles with leading auction and jewellery houses to academic posts. For ten years she was a jewellery specialist for Sotheby's and Christie's in London and Geneva, where she handled sales of global importance, including the private jewellery collection of HRH The Princess Margaret in 2006. She has since held roles as Professor of Jewellery at the Geneva University of Art and Design, Managing Director of the House of Gübelin, Head of Business Development at Gembridge, and most recently as the Senior Jewellery Curator at the Victoria and Albert Museum. *Precious* is her first book for general readers.

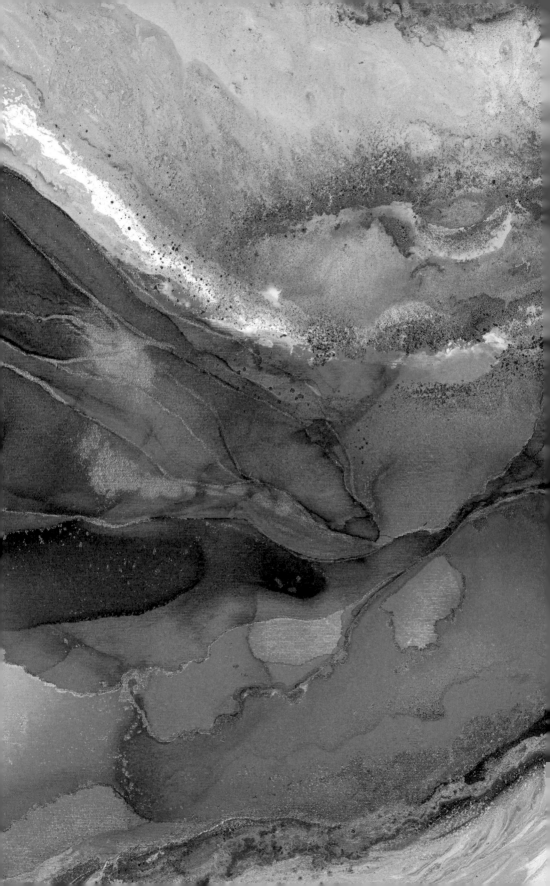